Approximating Perfection

Approximating Perfection

A Mathematician's Journey into the World of Mechanics

Leonid P. Lebedev
Michael J. Cloud

PRINCETON UNIVERSITY PRESS

PRINCETON AND OXFORD

Published by Princeton University Press, 41 William Street, Princeton, New Jersey 08540

In the United Kingdom: Princeton University Press, 3 Market Place, Woodstock, Oxfordshire OX20 1SY

Library of Congress Cataloging-in-Publication Data

Lebedev, L. P.

Approximating perfection: a mathematician's journey into the world of mechanics / Leonid P. Lebedev, Michael J. Cloud.

p. cm.

Includes biographical references and index.

ISBN 0-691-11726-8 (acid-free paper)

1. Mechanics, Analytic.   I. Cloud, Michael J.   II. Title.

QA805.L38 2004

531 — dc22   2003062201

British Library Catalog-in-Publication Data is available

The publisher would like to acknowledge the authors of this volume for providing the camera-ready copy from which this book was printed.

This book has been composed in Computer Modern

Printed on acid-free paper. ∞

www.pupress.princeton.edu

Printed in the United States of America

10 9 8 7 6 5 4 3 2 1

Contents

Preface

Although engineering textbooks once provided more breadth than they do today, few ever took the time to offer the reader a true perspective. We all know that myriad formulas are essential to engineering practice. However, modern textbooks have begun to allow formulas and procedural recipes to preoccupy the mind of the student. We have already reached a stage where proofs once deemed essential receive no mention whatsoever. The situation will undoubtedly worsen as computational methods demand an even greater share of the engineering curriculum. In some areas, we are rapidly nearing the point where even a passing familiarity with computational "recipes" will be deemed unnecessary: engineers will simply feed data into "canned" routines and receive immediate output. (It is likely that students will continue to welcome this prospect with open arms, until it finally dawns on them that the same task could be performed by someone who lacks a hard-earned engineering degree.) Unfortunately, all of this points to a diminishing grasp of just how complex real systems (industrial or otherwise) really are. One could argue that this is part of normal social progress: that a major goal of science should be our freedom from having to think too much. Why should the average person not be able to solve problems that surpassed the abilities of every true genius a century ago? But the argument quickly wears thin — anyone who engages in research and development activity, for instance, will certainly require a real understanding of things. Training in the use of rigid recipes may be appropriate for a fast-food cook, but not for the chef who will be expected to develop new dishes for persons having special dietary needs. The latter will have to learn a few things about chemistry, biology, even medicine, in order to function in a truly professional capacity.

This book is not a textbook on engineering mechanics, although it does contain topics from mechanics, the strength of materials, and elasticity. It considers the background behind mechanics, some aspects of calculus, and other portions of mathematics that play key roles in applications. The logic that underlies modeling in mechanics is its real emphasis. The book is, however, intended to be useful to anyone who must deal with modeling issues — even in such areas as biology. Students and experts alike may discover explanations that serve to justify routine actions, or that offer a better view of particular problems in their areas of interest.

The authors are grateful to Yu. P. Stepanenko of the Research Institute of Mechanics and Applied Mathematics at Rostov State University. Professor Stepanenko is an engineer and designer of measurement devices, and our fruitful discussions with him have resulted in many of the examples cited in this book. Edward Rothwell and Leonard Moriconi read large portions of the manuscript and provided valuable feedback. We are deeply indebted to our editor, Vickie Kearn, for several years' worth of help and encouragement. Thanks are also due to editorial assistant Alison Kalett, production editor Gail Schmitt, designer Lorraine Doneker, and copyeditor Anne Reifsnyder. Finally, Natasha Lebedeva and Beth Lannon-Cloud deserve thanks for their patience and support.

Chapter One

The Tools of Calculus

The complexity of Nature has led to the existence of various sciences that consider the same natural objects using different tools and approaches. An expert in the physics of solids may find it hard to communicate with an expert in the mechanics of solids; even between these closely related subjects we find significant differences in mathematical tools, terms, and viewpoints taken toward the objects of investigation. The physicist and engineer, however, do share a bit of common background: the tools of mathematical physics. These have evolved during the long history of our civilization.

The heart of any physical theory — say, mechanics or electrodynamics — is a collection of main ideas expressed in terms of some particular wording. The next layer consists of mathematical formulation of these main ideas. It is interesting to note that mathematical formulations can be both broader and narrower than word statements. Wording, especially if left somewhat "fuzzy," is often capable of a wider range of application because it skirts particular cases where additional restrictions would be imperative. On the other hand, mathematical studies often yield results and ideas that are important in practice — for example, system traits such as energy and entropy.

Among the most mathematical of the disciplines within physics is mechanics. At first glance other branches such as quantum mechanics or field theory might seem more sophisticated, but the influence of mechanics on the rest of physics has been profound. Its main ideas, although they reflect our everyday experience, are deep and complicated. The models of mechanics and the mathematical tools that have been developed for the solution of mechanical problems find application in many other mathematical sciences. The tools, in particular, are now regarded as an important part of mathematics as well. Indeed, the relationship between mechanics and mathematics has become so tight that it is possible to consider mechanics as a branch of mathematics (although much of mechanics lacks the formalized structure of pure mathematics).

In this book we shall explore the role of mathematics in the development of mechanics. The historical pattern of interaction between these two sciences may yield a glimpse into the future development of certain other fields of knowledge (e.g., biology) in which mathematical rigor currently plays a less fundamental role.

Our use of the term "mechanics" will include both "classical mechanics" and "continuum mechanics." The former treats problems in the statics and dynamics of rigid bodies, while the latter treats the motions, deformations, and stresses of bodies in cases where the details of atomic structure can be neglected. Thus mechanics describes the behavior of a great many familiar things. It can further account for the influence of phenomena previously not considered under the heading of mechanics: a mechanical body can exhibit magnetic properties, for example. Mechanicists developed simple but effective mathematical models of real objects used in engineering practice. These include models of beams, plates, shells, linearly elastic bodies, and ideal liquids. The development of mathematical tools for the solution of specific problems was done in parallel with this, and often by the very same persons. Early on, nobody tried to divide mathematics into pure and applied portions. There was a certain unity between mathematics and physics, and impressive amounts of progress were made on both fronts — even though the number of "research scientists" was considerably smaller than it is today. We know Sir Isaac Newton (1642–1727), for example, as both a great physicist and mathematician. The names Euler, Bernoulli, and Lagrange also summon images of first-rate mathematicians without whom much of continuum mechanics may not have survived. These historical icons, along with many others, developed new mathematical tools not only out of pure mathematical curiosity: they were also experts in applications, and knew what they needed in order to solve important practical problems. They understood the directions that mathematics needed to take during their lifetimes. Augustin-Louis Cauchy (1789–1857), whose name is encountered in any calculus textbook, introduced a key notion in continuum mechanics: that of the stress tensor. He elaborated it by exploiting the same tools and ideas he used to solve problems in pure mathematics. Many important branches of mathematics appeared in response to the very real needs of engineering and the other applied sciences.

Because a consideration of mathematical tools and how they are transformed in mechanical modeling is one of our goals, it makes sense to begin with a discussion of elementary calculus. Indeed, the ideas laid down in mechanical modeling are the same as those laid down in mathematics.

1.1 Is Mathematical Proof Necessary?

Among the most wonderful notions ever elaborated by mankind was that of number. This notion opened the door to comparisons that were previously impossible. Two men and two apples now had something in common: both of these sets could be mentally placed into correspondence with the same set of two fingers on one's hand. Many languages still contain a phrase along the lines of "... as many as the fingers on both of my hands." The sense of number is not exclusively human: some talented crows appear to have it as well. If such a crow is shown two sets of beans, one containing three beans and the other containing four, the bird can choose the larger set without error. It turns out that a crow cannot distinguish between sets of seven and eight beans, but even humans can normally grasp no more than six items at a glance and must count systematically after that. The notion of abstract number was brought into common use not so long ago (historically speaking). The necessity to count abstract quantities gave rise to methods called "algebra." Engagement in trade forced people to compare not only sets of discrete items, but tracts of land, quantities of liquid, and so forth. This led to geometry as the science by which continuous quantities could be compared. People elaborated the notions of area and volume; they had to compare pieces of land of various shapes, and therefore had to consider the main geometrical figures in the plane and in space and find their measures.

The ancient Egyptians and Babylonians could find the area of a rectangle and even a triangle. They had no formulas, so a solution was given as an *algorithm* describing which measurements must be taken and what must be done with the results. The methods of calculation were probably discovered during extensive testing: various areas were covered with seeds, which were then counted and the results compared with the output of some algorithm. Of course, this is only an assumption because all we have are the algorithms themselves, found on Babylonian clay tablets and similar sources; at that time nobody was interested in describing how they made such discoveries. Modern applied mathematical sciences also make extensive use of algorithms. Again, an algorithm is merely a precise description of the actions necessary to find some desired value. In this way, modern mathematics is similar to that of the Babylonians. It is possible that ancient mathematicians were even more careful than modern applied mathematicians: ancient algorithms were tested thoroughly, whereas today many algorithms are based largely on intuition (which is not so bad if the intuition is good).

The problem of comparison was the beginning of all mathematics. Of course, mathematics has evolved into a great many distant branches. But the common approach was developed by the ancient Greeks. The Greeks decided that it was necessary to lend support to the algorithms for the calculation of lengths, areas, and volumes known from even more ancient times. The word of an authority figure was no longer deemed sufficient: the Greeks also decided to require *proof.*

Interactions between civilized people are based on rules that all or most people consider to be valid. It is impossible to force people to do anything all the time; normal social relations are based on the act of convincing people to behave in certain ways. This in turn requires a convincing argument. This viewpoint culminated in ancient Greece where the laws of democracy forced people to learn the art of convincing others. The art became so important that certain persons began a systematic study of its main elements so that they could teach these to other people. Discussion itself was analyzed with the goal of figuring out how to win. Out of this came the main elements of logic and the standard modes of reasoning. Eventually some teachers left the practical realm of application of this knowledge and began to study Nature using the same approach. This led them to quite abstract notions. They developed an understanding of various effects, and tried to understand why things happen. The thinkers who had the most advanced tools, and who applied them in all situations, became known as philosophers. Among the most important realms was that which we now call *geometry*.

Ancient geometry combined the whole of the mathematics of its time. It was regarded as the ultimate abstraction and capable of explaining everything in Nature. Moreover, some people like Pythagoras expected that full knowledge of mathematical laws would give one power over the whole world. Numbers were thought to govern everything — learn their laws and you become king of the universe! This became a slogan for a certain direction of Greek philosophy that evolved into a kind of religion. The art of convincing argument was applied to the most abstract subject of the time: mathematics. Mathematics became full of abstract reasoning and was considered by the ancient Greeks to stand at the pinnacle of philosophy. This point of view came down to premodern times. In certain Russian universities prior to the October Revolution, for example, mathematics was relegated to departments of philosophy. This is why the Ph.D., the Doctor of Philosophy, is the title given to mathematicians and physicists.

To understand what the Greeks brought into mathematics, it is helpful to take a look at how their discussions were arranged. The dialogues of Socrates, as described by Plato, consisted of long chains of deductive reasoning. Each statement followed from the one before, and any chain of statements began with one that his adversary accepted as indisputable. We oversimplify Socrates' method a bit here, but the scheme we describe is the one that was introduced into mathematics. Mathematics became more than a set of practical algorithms; it was supplied with the tools one needed to argue that the algorithms were correct. The peak of this approach was presented in Euclid's *Elements*, which until recently was a standard text in geometry. Euclid (c. 300 B.C.) collected much of the geometry known in his time. His approach was copied by many other branches of mathematics and, when logical, by various other sciences.

What is the structure of the *Elements*? Euclid regarded certain central notions of geometry as self-evident. He did not try to define a point or a straight line, but only described their properties. These properties were taken as universally accepted. For example, a straight line gave the shortest distance between two points. The most evident of the basic assertions were called *axioms*, and the rest were called *postulates*.

With the axioms and postulates for the main elements (points, straight lines, and planes) in hand, the main figures (triangles, polygons, circles) can be introduced, and a study of the relations between these can begin. Not all "self-evident" statements remained so. Euclid's famous fifth postulate states that through a point outside a straight line on the plane it is possible to draw a parallel straight line, and that this line is unique. Later geometricians tried to prove this as a theorem. In the nineteenth century it was found to be an independent statement that might be true or untrue in real space — we still do not fully understand the real geometry of our space. So this postulate turned out to be nontrivial, and modern geometry regards it as an axiom for the so-called Euclidean geometry. But there are geometries with other sets of axioms. In Lobachevsky's geometry, through a point on a plane there are infinitely many straight lines parallel to a given one; in Riemannian geometry, it is taken for granted that on the plane there are *no* lines parallel to a given one. This results in some differences in theorems between the various branches of geometry. In non-Euclidean geometries, the interior angles of a triangle do not sum to 180°. A consequence is that we cannot precisely determine what qualifies as a straight line or a plane. We have only an image of a straight line as the path along which a ray of light reaches us from a distant star. But such a path is not straight: its course is altered by gravitation. So we can point to nothing in our environment that is a straight line. The same holds for the idea of a plane, notwithstanding the fact that children are told that the surface of the table in front of them is a portion of a plane.

The relations between geometrical figures were formulated as theorems, lemmas, corollaries, and so on, among which it is hard to establish strict distinctions. But all the central results were called *theorems*. These are formulated as certain conditions on the geometric figures, and the resulting consequences. All the theorems were *proved*, which meant that all the consequences were justified using the axioms and theorems proved earlier. This was a great achievement for a comparatively small group of people. In the *Elements*, even quadratic equations were solved through the use of geometrical transformations. This book was the standard of mathematicians for many years, and modern students who dislike similar logical constructions may consider themselves victims of the ancient Greeks.

Today, mathematics remains at the center of science. But it has developed along with everything else. Arguments considered rigorous by the ancient Greeks are sometimes considered to be flimsy nowadays. In 1899, an attempt was made to construct an absolutely rigorous geometry. In his famous *Foundations of Geometry*, David Hilbert (1862–1943) took the Greek formalization to an extreme level. Hilbert stated that there are to be undefined main elements — points, straight lines, and planes — and between these, some relations should be determined axiomatically. Then all the statements of geometry should be derived from these axioms independently of what one might regard as a point, line, or plane. The reader can view these terms in any fashion he or she wishes: students as points, desks as lines, and cars as planes. But despite a complete misunderstanding of these main elements, he or she should arrive at the same relations between them as would someone who had a better viewpoint (we say "better" because the notions are undefined, so nobody can have a perfect understanding). The geometry of Hilbert contains many axioms and is much more formal than that of Euclid. Later, failed attempts were made to prove consistency among the full system of geometrical axioms. The first attempt to axiomatize geometry spurred many mathematicians to try to axiomatize all of science. Mathematics was the pioneer of all rigorous study, and it has been said that the only real sciences are those that employ mathematics. (Should we believe such a statement?) Hilbert's book inspired other mathematicians to develop all of mathematics as an axiomatic–deductive science, where all facts are deduced from a few basic statements that have been formulated explicitly. Attempts have been made to introduce the axiomatic approach into mechanics and physics, but these have not been very successful. The totally axiomatic approach in mathematics was buried by Kurt Gödel (1906–1978), who proved two famous (and, in a sense, disappointing) theorems:

1. The *first inconsistency theorem* states that any science based on the axiomatic approach contains undecidable statements — theorems that could not be proved or disproved. Therefore such statements can be taken as independent axioms of any version of the theory: positive or negative. This means that any axiom system is incomplete if we wish to find all the relations between some elements for which the axioms were formulated.

2. The *second inconsistency theorem* states that it is impossible to show consistency of a set of axioms using the formalization of the system itself; only by using a stronger system of axioms can we establish such consistency. This means that even the simplest axiom system describing arithmetic cannot be checked for consistency using only the tools of arithmetic.

This did not mean the end of mathematics. But it did bring some indefiniteness into the strictest of the sciences. The way in which the ancient Greeks proved mathematical statements remains the most reliable method available, and students will continue to study proofs of the Pythagorean theorem for years to come.

1.2 Abstraction, Understanding, Infinity

A small boy who announces that he fails to understand the concept of "one" is not necessarily being silly. There is no such thing as "one" in Nature. We can point to one finger or one pie, of course, but not to "one" as an independent entity. The concept of "one" is quite fluid: we can use it to refer to a whole pie or to a single piece of that pie. So the boy may have good reason to be perplexed. It is probably fair to say that we become accustomed to abstract notions more than we actually understand them. Two high school trigonometry students may exhibit identical skill levels and yet disagree profoundly as to whether the subject is understandable. This sort of difference is essentially psychological, and in many cases relates directly to how long we have received exposure to an abstract concept. Young persons seem to be especially well equipped to accept new notions and to use them with ease. As we get older we seem to require much more time to decide that something is clear, understandable, and easy.

One abstract notion that takes a while to accept is that of infinity. We overuse the word "infinite," even applying it when we are angry about having to stand in line too long. But infinity is a troublesome concept: we are at a loss to really visualize it, yet seem to be able to vaguely intuit more than one version of it. One could write volumes about how to imagine infinity. However, the thicker the book, the less clear everything would become; our understanding should spring from the simplest possible ideas, and proceed systematically to the more complex.

Centuries ago, there were attempts to treat infinite objects in the same way as finite geometrical objects. In this way, the properties of finite objects were attributed to infinite objects. An "actual infinity" was introduced and treated as though it were an ordinary number. Many interesting arguments were constructed. For example, one could "prove" that any sector of a plane (i.e., the portion lying between two rays emanating from a common point) is larger than any strip (i.e., a portion lying between two parallel lines). The argument was simple: an infinite number of strips will be needed to cover the plane, whereas a finite number of sectors will suffice. So the area of the sector must be larger. This and many similar "proofs" introduced paradoxes into mathematics and forced people to reconsider the concept of infinity. What finally remained was a "potential infinity" that could be only approached using processes involving finite shapes or numbers.

Let us see how infinity enters elementary mathematics. We should take the simplest object we can and try to understand how infinity comes into it. This object is the set of positive integers. It contains no largest element: any positive integer can be increased by one to yield a new positive integer, so we say that the positive integers stretch on to infinity. This is an infinite set with the simplest structure. We can compare the sizes of two sets by using the idea of a one-to-one correspondence. We say that two sets have the *same cardinality* if they can be placed in such a correspondence, regardless of the nature of the elements themselves. The positive integers and its various subsets are useful for such comparisons. This is essentially what we do when we count the elements of a set: we say that a set has n elements if we can pair each element uniquely with an integer from the set $\{1, 2, 3, \ldots, n\}$. When no value of n suffices for this, we say that our set is infinite. The set of positive integers has the least possible cardinality of all infinite sets, and any set with the same cardinality is said to be *countable*. Thus, each member of a countable set can be labeled with a unique positive integer. It is clear that if we put two countable sets A and B together (i.e., form the *union* of the sets), we obtain a new countable set. The elements of the new set can be counted as follows: the first member of A is paired with the integer 1, the first member of B is paired with the integer 2, the second member of A is paired with the integer 3, the second member of B is paired with the integer 4, and so on. Thus, from the viewpoint of cardinality, the union of A and B is no larger than either of these individual sets. The same thing happens when we consider the union of any finite number of countable sets: the result is countable. Moreover, the union of countably many sets is always countable. (The reader could try to demonstrate this by showing how the elements of the union can be renumbered so that each is paired uniquely with an integer.) From this, it follows that the set of all rational numbers is countable. From one perspective, the rational points cover the number line densely, so that in any neighborhood of any point we find infinitely many rational points; however, in the sense stated above, the rationals are no more numerous than the positive integers — which are quite rare on the same axis. Of course, this is a bit bothersome. Another interesting question is whether there exists a set with cardinality higher than the positive integers. Mathematicians have shown that the set of points contained in the interval $[0, 1]$ is of this type: these points are too numerous to be paired with the positive integers. We say that $[0, 1]$ has the cardinality of the *continuum*. It turns out that the same cardinality is shared by the entire number line, the plane, and three-dimensional space as well. Thus we have come to consider another kind of infinity. We shall not touch on infinities of even higher order, though they exist; the curious reader can consult a textbook on set theory for more information. We should, however, ask whether sets exist with cardinalities intermediate between those of the positive integers and the continuum. The assumption that they do, it turns out, is a fully independent axiom of arithmetic. The stipulation of this axiom (or lack thereof) will not affect the results on which we so often depend. This is a confirmation of Gödel's famous theorem, because it demonstrates the incompleteness of the axioms of arithmetic — the simplest (and seemingly nicest) axiom system in mathematics.

Our easy familiarity with the positive integers leads us to think that we should be able to perform countably many actions ourselves. Indeed, we effectively do this when we sum an infinite series via a limit passage. After becoming bored with such things, mathematicians decided to attempt actions as numerous as the points of the continuum. But this led to serious paradoxes. It was found, for example, that one could carve up a sphere of unit radius into a great many pieces, alter their shapes without changing their volumes, and then collect the new pieces into a sphere of radius 1,000,000 with no holes in the latter (the Banach–Tarski Paradox[1])! Hence the problem of infinity was considered more carefully. This problem is the central point of calculus and other parts of analysis that relate to differentiation and integration, but there, limit passages are presented in such a way that only finite quantities are involved at each step. This explains why modern students can quickly feel secure in their necessary dealings with the subject.

1.3 Irrational Numbers

When we purchase 1 meter of cloth, we may actually receive 1.005 meters. Of course, we probably will not care; for our purposes the tiny bit of extra cloth will not matter. We might even refer to two pieces of cloth having

---

[1]See *The Banach–Tarski Paradox* by Stan Wagon (Cambridge: Cambridge University Press, 1993).

equal widths, with lengths of 1 m and 1.005 m, respectively, as being of "equal" size. This kind of fuzzy thinking has its place in everyday life. But it was also characteristic of very ancient mathematics.

The Greeks refused to accept this approach. In their view equality was *absolute*. Two triangles were considered equal only when the first could be exactly superimposed upon the second; any difference, no matter how small, made them unequal. Now, it may be hard to imagine a mathematics based on "approximate" equalities, but things could have developed that way. We would have lost algebra and the strict results of geometry, but might have gained something else. Let us avoid speculation, however, and consider what actually did happen instead.

In geometry, one of the simplest and most crucial problems was the measurement of the length of a straight segment. The length of a given segment should be compared with the length of a standard segment that can be said to possess unit length. Any physics student is aware that the meter length was determined by the length of a standard metal rod kept under lock and key somewhere by a standards organization. We measure everything in terms of the length of this rod. But although the rod is maintained under controlled environmental conditions, its length will vary a bit due to tiny changes in air temperature, pressure, and so on, in its holding chamber. These changes may be small but they are almost certainly not zero, and are definitely an issue when we are trying to determine whether two lengths are absolutely equal. Even subtle changes in electromagnetic fields caused by solar activity will alter the length of the standard meter bar.

Retaining the viewpoint of the ancient Greeks who introduced absolute equality, we may or may not even be able to find two moments in time when the lengths of this same bar will be equal.

Thus, the proudly absolute Greeks proposed a way in which the length of an arbitrary segment could be measured in terms of a unit-length standard segment. Our modern adaptation is as follows. We place the standard segment next to the unknown segment with two of the ends flush. If the unknown segment is longer, then we mark the position of the other endpoint of the standard segment on the unknown segment, and repeat this procedure until the portion of the unknown segment that remains is less than the length of the standard segment. In this way, we find the integer part of the unknown length. We next divide the standard segment into several equal parts, say ten, and repeat the procedure using one-tenth of the standard segment. We can continue to subdivide the standard, and in this way determine the unknown length to any desired degree of accuracy.

What the Greek mathematicians did was actually a bit different. They first supposed it possible to divide a standard segment into several equal parts, say n parts, in such a way that the unknown segment is composed of an integer number of such parts. If this integer number is m, then in modern notation the unknown length is m/n. At first, the Greeks were sure this would always work. Their certainty was rooted in their belief that the structure of the world was ideal (and many modern scientists share this view). It is interesting to note that the Greek philosophers held notions akin to our present knowledge of the atomic structure of materials. They did, however, have a different idea of what an atom would look like if one could be seen. Atoms were thought to be objects that completely filled in the exterior framework of a material body. Because all atoms of a certain material had to be identical, and these had to fill in the body completely without gaps, atoms were supposed to take some ideal form — possibly that of a perfect polyhedron. This formed the basis of a mathematical/religious belief that integer numbers govern the world. And so the ancient mathematicians thought they could measure the proportions of any perfect figure through the use of integers only.

This belief persisted until it failed for the square. If a square has sides of length 1, a quick calculation using the Pythagorean theorem shows that the diagonal has a length, in modern terms, of $\sqrt{2}$. If the above method of measuring lengths were valid, this length would have to equal m/n for some integers m, n. We will demonstrate that this cannot be the case. Let us first assume that m and n have no common integer factors: that is, that all factors common to the numerator and denominator of the fraction m/n have been cancelled already. Let us now suppose that there are integers m, n such that

$$\sqrt{2} = m/n.$$

Squaring both sides of the equality, we get

$$2 = m^2/n^2,$$

or

$$m^2 = 2n^2.$$

Now, m^2 cannot have 2 as a divisor unless m does also. So there must be another integer p such that $m = 2p$. Substituting this into the last equality, we have

$$4p^2 = 2n^2,$$

or

$$n^2 = 2p^2.$$

But this means that n also has 2 as a divisor, and can therefore be expressed as $n = 2q$ for some integer q. This, of course, contradicts our assumption that m and n have no common integer factors. Hence, $\sqrt{2}$ cannot be presented in the form of a ratio m/n in which m, n are integers; we now say that $\sqrt{2}$ is not a *rational number*. This spelled the end for some of the old Greek ideals, but brought into play a new class of numbers called *irrational numbers*.

What is an irrational number? The symbol $\sqrt{2}$ is nothing more than a label unless we can say precisely what this number is. And this we actually cannot do, but we can calculate the value of $\sqrt{2}$ to within any desired accuracy. We can find as many decimal digits of this number as we wish, but can never state its exact value except by giving the label $\sqrt{2}$. Irrational numbers are not exceptional cases. In higher mathematics, it is shown that the situation is just the opposite: rational numbers are the exceptions, and in a certain technical sense the number of rationals is negligible in comparison with the number of irrationals. So we should elaborate on how to describe an irrational number precisely. There are two ways of doing this. The first is simply to label the number. This is possible for some important irrational numbers such as π and e, but not for all of them. The second is to describe how to pin down an irrational number more and more precisely. Thus we can say that an irrational number is the result of some infinite process of approximation. In this way we arrive at the notion of *limit*. The Greeks used this notion in an intuitive manner, but the strict version had to wait until much later — when it could form the backdrop for the calculus.

1.4 What Is a Limit?

Let us pursue the question of how irrational numbers are defined. This leads us to the more general idea of limit. We will need the notion of distance from geometry. We introduce a horizontal line (*axis*) on which we can mark a zero point, a unit distance away from zero, and a positive direction (say, to the right). Using these, we can mark a point corresponding to a rational number m/n. This is the point whose distance from zero is $|m/n|$; it falls to the right of zero if m/n is positive or to the left of zero if m/n is negative.

In this way, we can mark enormously many points on the numerical axis, but some points (e.g., $\sqrt{2}$) must remain unmarked. Note that we have in mind a nice picture of a straight line. But remember that we really do not know what a straight line is: we only know some of its properties. In geometry it is often said that a straight line is a portion of a circle having infinite radius. This notion can be bothersome if we try to envision the full circle; it is reasonable, however, if we concentrate only on a certain portion of interest, and the idea does work nicely in doing calculations! Another question worth considering is whether we could ever draw even a segment of a truly straight line, or come up with some other suitable material representation. If we keep in mind that all materials are made up of separate atoms, then it becomes hard to imagine constructing a continuous straight segment. If we add that some modern physical theories regard space as having a cell structure, then drawing a straight line becomes more than problematic. However, idealization is still a convenient way to investigate relations between real-world objects. To precisely define the concept of an irrational number, we can apply the idea of successive approximations as introduced in the previous section. We will modify that idea as follows.

To locate any real number a, whether irrational or rational, we begin by bracketing it between two points on either side. Let us write

$$a_0^- \leq a \leq a_0^+,$$

where the *lower* and *upper bounds* $a_0^\pm$ are rational. We can refine our approximation of a by giving another rational pair $a_1^\pm$, between which a is squeezed more tightly: that is, we want

$$a_1^- \leq a \leq a_1^+,$$

where $a_0^- \leq a_1^-$ and $a_1^+ \leq a_0^+$. Repeating this process, we can bracket a more and more precisely, and thereby think of approximating it more and more closely. Our construction is such that the kth approximating segment $[a_k^-, a_k^+]$ lies within all the previous segments $[a_i^-, a_i^+]$, $i = 0, 1, \ldots, k-1$, and the lengths $|a_k^+ - a_k^-|$ decrease toward zero as k increases indefinitely.

It seems evident that some point on the axis will belong to each of the approximating segments $[a_k^-, a_k^+]$, and that this point will be unique. Unfortunately, because we know little more than the ancient Greeks knew about straight lines, we do not know anything about the "local structure" of the line and must therefore take this "evident" statement for granted. (The reader might consider proposing some alternatives: perhaps there are no such points, or two, or infinitely many?) This is an example of an axiom: a statement that cannot be proved but which can be accepted so that argumentation can proceed. So we hereby introduce an axiom of nested segments: there is a unique point that belongs to each of a sequence of nested segments whose lengths tend ever toward zero.

Despite the fact that things are looking evident and even fairly scientific now, we have managed to slip in another fuzzy notion. What does the phrase "tend toward zero" mean? A precise definition was proposed by Cauchy, the French mathematician so prolific that the flow of his papers overwhelmed the mathematical journals of his time (he eventually had to publish his own journal). When we hear such impressive sounding names we are prone to imagine persons whose lives were led in a sophisticated and academic manner, and who from time to time published important results that might immediately come to be called Cauchy's theorem or Lagrange's theorem. But in ordinary life, great men sometimes behave as badly as anyone else. Cauchy's results were so numerous that, as the story goes, another of his great contemporaries characterized Cauchy as having "mathematical diarrhea." In return, Cauchy claimed that his contemporary had a case of "mathematical constipation." We shall meet Cauchy again on future pages; he was truly great and originated many important ideas. One of these was quite nontrivial: how to describe infinite processes in terms of finite arguments — again, the notion of limit. Despite our desire to be extremely advanced in comparison with other animals, we find ourselves quite restricted when we attempt to do anything exactly. We cannot tell whether a bag contains eight or nine apples without counting them, just as a crow could not (there are exceptional persons such as that portrayed by the main character of the movie *Rain Man*, but these persons merely possess higher ceilings for making immediate counts). We definitely cannot perform infinitely many operations even if we declare that we can. So we need a way to confirm that all has gone well in a process having an infinite number of steps, even if we can carry out only finitely many. This is the central point of approximation theory, and thus the theory of limits.

We consider the simple notion of sequence limit. The notion of sequence is well understood. For example, a sequence with terms $x_n = 1/n$ looks like

$$1, \quad 1/2, \quad 1/3, \quad \ldots, \quad 1/n, \quad \ldots.$$

If we consider the term numbered $n = 100$, we get $x_{100} = 1/100$. For $n = 1000$ we get $1/1000$, which is closer to zero; for $n = 1000000$ we get $1/1000000$, which is even closer. The higher the index n, the closer the sequence term is to zero. Thus zero plays a special role for this sequence and is called the *limit*. In our example, the sequence terms decrease monotonically, but this does not always happen, and we can phrase the limit idea more generally by saying something like "the greater the index, the closer the terms get to the limit." Cauchy proposed the following definition.

Definition 1.1 The number a is the *limit* of the sequence $\{x_n\}$ as $n \to \infty$ if for any $\varepsilon > 0$ there is a number N, dependent on ε, such that whenever $n > N$, we get $|x_n - a| < \varepsilon$. A sequence having a limit is called *convergent*, and one having no limit is called *nonconvergent*. If a is the limit of $\{x_n\}$, then we write

$$a = \lim_{n \to \infty} x_n,$$

or $x_n \to a$ as $n \to \infty$.

Note that if we consider a stationary sequence — that is, a sequence whose elements $x_n = a$ for all n — then by definition a is the limit. This supports our view of the limit as a value approached more and more closely as the index n increases.

Let us carefully consider the limit definition from several points of view, because it is the central definition of the calculus and of continuum physics. First, there is the requirement "for any $\varepsilon > 0$". The word "any" is dangerous in mathematics. In this context, it means that if we are to verify that a sequence has limit a, we must try *all* positive values of ε without exception; for each $\varepsilon > 0$, we must be able to find the number N depending on ε mentioned next.

At first glance, we appear to have gained nothing: we started with an infinite process of either convergence or nonconvergence to a limit, and have replaced it by another, apparently infinite, process that forces us to seek a suitable N for every possible positive number ε. Note that it can be described as follows. We take some sequence of values ε_k tending toward zero, and for each of these find an $N(\varepsilon_k)$. It is clear that if we find $N(\varepsilon_1)$ then this N can be taken for all $\varepsilon > \varepsilon_1$. Thus we need not consider the continuous set of all $\varepsilon > 0$. Moreover, although the definition states that we should verify the property for all ε, our serious work begins with those that are close to zero. But the new process need not be infinite: we can find an analytical dependence of N on ε by solving the inequality

$$|x_n - a| < \varepsilon$$

with respect to the integer n, and then see whether the solution region contains some interval $[N, \infty)$. If so, then a is the limit of the sequence. If there is an $\varepsilon > 0$ for which there is no such N, then the sequence has no limit. Thus our task does not involve infinitely many checks, but rather the solution of an inequality. This is how the result of an infinite process can be verified through a finite number of actions.

Note that the definition of sequence limit does not give us a way to find the limit a. We must produce a candidate ourselves before we can check it using the definition.

Let us reformulate the definition in a more geometrical way. We rewrite the above inequality in the form

$$|x - a| < \varepsilon.$$

The set of all x that satisfy this is $(a - \varepsilon, a + \varepsilon)$, which is called the ε-*neighborhood* of a on the numerical axis. So, geometrically, when we find the solution $[N, \infty)$ of the above inequality, we find the number N beginning with which all the terms lie in the ε-neighborhood of a. That is, if we remove all terms of the sequence whose indexes are less than or equal to N, then the whole remaining "tail" of the sequence must fall within the ε-neighborhood of a. The number a is the limit if we can do this for $\varepsilon > 0$ without exception. The "tail" idea can be made clearer if we think of the sequence $\{x_n\}$ as a function given for an integer argument n. Figure 1.1 serves to illustrate this viewpoint. The picture extends infinitely far to the right ($n \to \infty$), and we see that convergence or nonconvergence of the sequence depends on what happens to the far right. If we were to drop some of the first terms (even the first few billions of terms, which would still be just the "first few" in comparison with the infinitely many terms present), there would be no effect on convergence or nonconvergence.

We will not consider all the properties of sequence limits, but would like to add one important fact. When a sequence is convergent, then it has one *and only one* limit. This is referred to as *uniqueness* of the limit, and is a property that can be established as a theorem based on the limit definition.

Now our formulation of the axiom for nested segments is exact, and we can return to the problem of defining an irrational number.

We represented each irrational number through some limiting process using nested segments (containing the number) whose lengths tend to zero. In

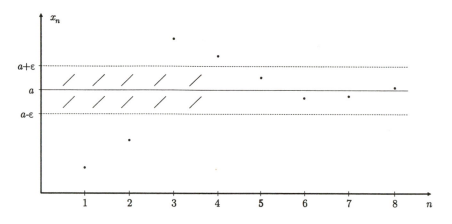

Figure 1.1 One view of sequence convergence. All terms of the sequence after x_4 are inside the band of thickness 2ε (between $a - \varepsilon$ and $a + \varepsilon$), the ε-neighborhood of a. In order to verify that a is actually the limit of $\{x_n\}$, we must verify that a similar picture holds for any positive ε.

fact we can do the same thing for any rational number x, by considering zero-length nested segments $[x, x]$. So each real number, whether rational or irrational, can be defined in a similar mode. Of course, this sort of academic "stuff" would be useless if the arithmetic of such numbers were to give results that disagreed with those of ordinary arithmetic. It turns out that we can employ the usual four arithmetic operations in the same way as we do for fractions. When we add two numbers, whether rational or irrational, the nested approximating segments for these define the ends of segments containing the sum. These segments turn out to be nested as well, and their lengths tend to zero along with the lengths of the original segments that bracketed the operands. In this way we get limit points that define the sum. This can clearly be done for subtraction, multiplication, and division as well. The details can be found in more advanced books. We should also note that our description of an irrational number is not unique: other valid interpretations have been invented, and all have been shown to be equivalent. But the important fact is that we can treat rational and irrational numbers similarly when we do arithmetic and other operations.

1.5 Series

The notion of limit is central in continuum physics. It brings us various mathematical models, at the core of which lie differential or integral equations. These relations are usually derived from geometrical considerations involving infinitesimally small portions of objects. Together with some additional conditions given on the boundary of the object, they form so-called boundary value problems. Because the solutions are frequently given in terms of *series*, we would like to examine this important concept. A series is an "infinite sum," commonly denoted as

$$\sum_{n=1}^{\infty} a_n,$$

which simply means that we must perform the addition

$$a_1 + a_2 + a_3 + a_4 + \cdots$$

step by step to infinity. It is clear that this involves a limit passage. Elementary math students are frightened of series, sometimes refusing to believe that any infinite sum of positive numbers could turn out to be finite. The ancient mathematician-philosophers expressed similar reservations: the famous Zeno (c. 495–430 B.C.), while considering the nature of space and motion, constructed a well-known paradox about the swift runner Achilles, who could never catch a turtle.

The proof was as follows. Between Achilles and the turtle, there is some distance. By the time Achilles covers this distance, the turtle has moved ahead by some distance. By the time Achilles covers this new distance, the turtle has moved ahead by yet another distance. The resulting process of trying to catch the turtle never ends. For Zeno, it was evident that a sum of finite positive quantities must be infinite, which meant that the time necessary for Achilles to catch the turtle must be infinite. Here, Zeno was led to consider a sum of time periods of the form

$$t_1 + t_2 + t_3 + t_4 + \cdots,$$

which, again, he regarded as obviously infinite.

We shall see that certain sums of this type are infinite, but others can be interpreted in a limit sense. The latter are useful in practice. A common but sometimes overlooked example is the ordinary decimal representation of the number π:

$$\pi = 3.1415926535897\ldots.$$

This can be viewed as the infinite sum

$$3 + 0.1 + 0.04 + 0.001 + 0.0005 + 0.00009 + 0.000002 + \cdots.$$

We have all used some approximation of π, truncating the series somewhere (usually at 3.14) with full awareness that we would have to include "all of the terms" to get the "exact result." So in some sense it must be possible to "sum" an infinite number of terms and obtain a finite answer. (The Greeks did not have a positional notation for decimal numbers, hence this argument would probably not have convinced them. This way of representing numbers was invented later by the Arabs.) Let us consider what it means to sum a series. Consider a finite sum

$$S_n = a_1 + a_2 + \cdots + a_n.$$

This can approximate the full sum of a series, just as 3.14 can approximate π. Increasing the number of terms by one, we get

$$S_{n+1} = a_1 + a_2 + \cdots + a_n + a_{n+1}.$$

In this way, we get a sequence

$$\{S_n\} = (S_1, S_2, S_3, \ldots, S_n, \ldots)$$

that is no different from the sequences we considered in the previous section. Hence we may consider the problem of existence of the limit for such a sequence. If this limit

$$S = \lim_{n \to \infty} S_n$$

exists, then it is called the sum of the series

$$\sum_{n=1}^{\infty} a_n,$$

and the same symbol $\sum_{n=1}^{\infty} a_n$ is used to denote the sum S as well. In this case, the series is called *convergent*. If the sequence S_n has no limit, then the series is called *nonconvergent*, but the symbol for an infinite sum is still written down formally. In higher mathematics, it is found that there are nonconvergent series whose finite sums S_n yield good approximations to certain functions for which formal infinite sums have been derived. Such series are called *asymptotic expansions*. A well-known example is the Taylor series for a nonanalytic function.

Having seen that irrational numbers can be represented as convergent series, we understand that the series idea is useful and convenient. But this is only the beginning. Many functions can be also represented as convergent series. An example is the expansion into a power series of $\sin x$:

$$\sin x = x - \frac{x^3}{3!} + \frac{x^5}{5!} - \frac{x^7}{7!} + \cdots .$$

With this, we can calculate the value of $\sin x$ for small x to good accuracy by using a finite number of terms. Moreover, for $|x| < 1$, the error is less than the last term at which we break off the sum. If we calculate the sum of the first three terms for $x = 0.1$, which is $0.1 - 0.001/6 + 0.00001/120$, then we get $\sin 0.1$ with ten digits of precision.

In past centuries, this and similar series were used to produce tables of functions like $\sin x$, $\cos x$, and so on. The same process is implemented by modern electronic calculators. The speed of computers has made it easier to include more and more terms in an approximation, somewhat alleviating the need to justify the accuracy of each calculation on a theoretical basis. However, the fundamental problem of justifying series convergence remains. We must understand that no result for a finite number of terms can justify the convergence of a series — the situation is the same as with the limit of a sequence.

A few series can be calculated exactly. An example is the series of terms of a decreasing geometric progression:

$$b, bq, bq^2, bq^3, \ldots, bq^n, \ldots \qquad (|q| < 1).$$

The first n terms sum according to the formula

$$S_n = b\frac{1 - q^{n+1}}{1 - q}.$$

It is clear that $\lim_{n\to\infty} q^n = 0$ (the reader can either prove this or become convinced by evaluating a few terms on a calculator), and thus,

$$b + bq + bq^2 + \cdots + bq^n + \cdots = \lim_{n\to\infty} S_n = \lim_{n\to\infty} b\frac{1 - q^{n+1}}{1 - q} = \frac{b}{1 - q}.$$

This series plays a key role in the theory of series convergence, because it leads to the following comparison test.

Theorem 1.2 If a series $\sum_{n=1}^{\infty} a_n$ with positive terms a_n is convergent, then a series $\sum_{n=1}^{\infty} b_n$ with terms such that $|b_n| \leq a_n$ is also convergent. If $a_n \leq b_n$, and the series $\sum_{n=1}^{\infty} a_n$ is nonconvergent, then the series $\sum_{n=1}^{\infty} b_n$ is also nonconvergent.

The geometric progression is widely used to determine whether a series is convergent or nonconvergent. In the case of convergence, it also affords an estimate of the error due to finite approximation of the series, and can give an indication of the convergence rate. Some of the most capable mathematicians in history have devised series convergence tests. According to Cauchy, we can investigate the behavior of $\sum_{n=1}^{\infty} a_n$ by examining the behavior of $\sqrt[n]{|a_n|}$ as $n \to \infty$. If the limit is less than 1, then the series is convergent; if it is greater than 1, the series is nonconvergent; if it is equal to 1, no conclusion can be drawn. A similar test due to Dirichlet[2] has us take the limit of the ratio a_{n+1}/a_n. The conclusions regarding convergence are the same as in the previous test.

---

[2]Peter Gustav Lejeune Dirichlet (1805–1859).

The reader can verify that these tests do not yield a definitive result for the series

$$1 + \frac{1}{2} + \frac{1}{3} + \frac{1}{4} + \cdots, \qquad 1 + \frac{1}{2^2} + \frac{1}{3^2} + \frac{1}{4^2} + \cdots,$$

which are nonconvergent and convergent, respectively. But convergence of the latter is slow and we must retain many terms to get good accuracy.

We will meet further series in the following sections.

1.6 Function Continuity

A sequence can be regarded as a function whose argument runs through the positive integers and whose values are the sequence terms. That is, we may regard $\{x_n\}$ as the set of values of a function $f(n)$ by taking $f(n) = x_n$. But we know that many other functions exist whose arguments can vary over any set on the real axis. We shall now extend the limit concept to functions in general. It is here that the important notion of continuity arises.

Let us briefly review the idea of a real valued function. It is worth doing this because, in continuum mechanics, we will need to consider its generalizations to functions taking vectorial and tensorial values, and even to functions having other functions as arguments; the latter are called *functionals* if their values are real or complex numbers, and *mappings* (or *operators*) if their values are functions or other mathematical entities. We could say the following:

> A function f is a correspondence between two numerical sets X and Y, under which each element of X is paired with at most one element of Y. The set of all $x \in X$ that are paired with some $y \in Y$ is called the *domain* of f and is denoted $D(f)$. The set of all $y \in Y$ that correspond to some $x \in D(f)$ is called the *range* of f and is denoted $R(f)$.

In high school mathematics, we learn about functions for which X is a subset of the real axis, but more general sets of arguments x are common in higher mathematics. The notion of function reflects our broad experience with such relations between many characteristics of objects in the real world.

As usual, we shall not cover all the material given in standard textbooks on calculus, but we will consider the main properties of the tools of calculus — derivatives and integrals — that will be needed in what follows.

We are interested in what happens to the values of a function $y = f(x)$ at a point $x = x_0$. Our experience with graphing simple functions in the plane suggests that most graphs are constructed using continuous curves. These curves may be put together in such a way that the resulting graph has some points at which unpleasant things happen: the graph can approach infinity or possess a sharp break. It is natural to refer to the portions of the graph

away from such points as being "continuous." However, this cannot serve
as a definition of continuity because some functions are so complex that we
cannot draw their graphs. We would first like to understand what happens
to the values of a function $f(x)$ as x gets closer and closer to some point
x_0. An important case is that in which $f(x)$ is not determined at x_0. Let
us bring in the notion of sequence limit as discussed in the previous section.
Thus, we consider a sequence $\{x_n\}$ from the domain of $f(x)$, which tends
to x_0 but does not take the value x_0. The corresponding sequence of values
$\{f(x_n)\}$ may or may not have a limit as $n \to \infty$. If not, we say that $f(x)$
is discontinuous at x_0. Suppose $\{f(x_n)\}$ has a limit y_0. Is this enough to
decide that everything is all right at x_0? The answer is no; it is possible that
we have chosen a very special sequence. A good example of this is afforded
by the function $\sin(1/x)$ considered near $x_0 = 0$. Taking $x_n = 1/(\pi n)$, we
get $f(x_n) = 0$ for each n, hence the limit of the sequence of function values
is zero; but taking $x_n = 1/(2\pi n + \pi/2)$, we get $f(x_n) = 1$ for each n, and
this time the limit is one. Here, two sequences that both tend to zero are
mapped by the function into sequences that converge to different limits.

It makes sense to introduce the following notion of the limit of $y = f(x)$
at a point x_0.

Definition 1.3 A number y_0 is called the *limit* of a function $y = f(x)$ as x
tends to x_0 if, for any sequence $\{x_n\}$ having limit x_0 and such that $x_n \neq x_0$,
the number y_0 is the limit of the sequence $\{f(x_n)\}$. This is written as

$$y_0 = \lim_{x \to x_0} f(x).$$

Here, when we refer to "any sequence" we refer to all sequences $\{x_n\}$ from
the domain of $f(x)$ having the above properties.

This definition looks nice except for one thing: we must verify the desired
property for *every* sequence tending to x_0. Such an infinite task is beyond the
abilities of anyone. But Cauchy saved the day again by introducing another
definition of function limit. Cauchy's definition turns out to be equivalent
to that above, and reminds us of the definition of sequence limit.

Definition 1.4 A number y_0 is the *limit* of a function $y = f(x)$ as x tends
to x_0 if for any $\varepsilon > 0$ there is a $\delta > 0$ (dependent on ε) such that for any x
from the δ-neighborhood of x_0, except x_0 itself, the inequality

$$|f(x) - y_0| < \varepsilon \tag{1.1}$$

holds.

The required condition must be fulfilled for any positive ε; have we therefore simply replaced one infinite task with another? No. Our previous experience with sequence limits shows that we can solve the inequality (1.1) with

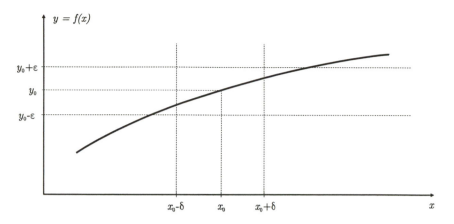

Figure 1.2 The ε–δ definition of function continuity: y_0 is the limit of $f(x)$ at x_0, if appointing a horizontal band of width 2ε about y_0, we can always find a vertical band centered at x_0 that is 2δ wide and such that all points of the graph of $y = f(x)$ (except the one corresponding to $x = x_0$) that lie in the vertical band are simultaneously in the horizontal band. This procedure should be possible for any positive ε.

respect to x for an arbitrary positive ε, and we can show that the solution domain includes the two intervals $(x_0 - \delta, x_0)$ and $(x_0, x_0 + \delta)$, where $\delta > 0$ depends on ε in some fashion.

This is presented in Figure 1.2. At point x_0, the function $f(x)$ can be defined or undefined. If it is defined and if we find that the limit at this point is equal to the defined value, that is, if

$$f(x_0) = \lim_{x \to x_0} f(x),$$

then we say that $f(x)$ is *continuous* at x_0.

We say that $f(x)$ is continuous in an open interval (a, b) if it is continuous at each point of the interval.

Continuous functions play an important role in the description of real processes and objects. Both continuous and discrete situations can be described. For example, objects are sometimes described by parameters that take on discrete values; however, we may choose to employ continuous functions as an approximation to this situation. On this idea, we base all theories of motion of material bodies — bodies that actually consist of separate atoms.

1.7 How to Measure Length

Perhaps the reader has already studied much more mathematics than we are discussing in this book. But it is probable that his or her background

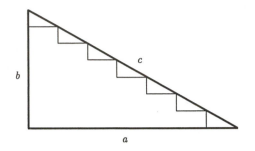

Figure 1.3 An inaccurate "approximation" of the length of a line (hypotenuse, of the length c, of a right triangle) by a polygonal curve having length $a + b$, the sum of the legs.

was slanted toward immediate applications rather than toward the theory needed to understand why mathematical "tools" work properly. We would therefore like to reconsider some topics and discuss some ideas that reside in the background of mathematics and mechanics. One important problem is this: how should we measure the length of a curve? The answer seems simple at first; we can use a tape measure or similar device. But beneath the surface of apparently simple questions we often find complex issues that require sophisticated tools for complete understanding. The calculation of length along a curve will evidently require some suitable method of approximating the desired value, followed by application of a limit passage. Here, though, even the first step — the construction of a first approximation — is not obvious. Our discussion of the problem will follow that of Henri Lebesgue (1875–1941), a mathematician known for developing a more abstract theory of integration as well as some important ideas in mechanics.

Thus, we turn to a problem of finding the length of a portion of a curve. We seek an exact result, hence ideas like placing a thread along the curve and then straightening it out against a ruler will not suffice. (It is unclear, for example, how precisely we could superimpose even a fine thread onto our curve, or what might happen to the length of the thread when we straighten it out.) So we must propose a purely geometrical mode of measurement that relies only on what we know about measuring straight segments. Because we cannot superimpose a straight segment onto a curve, our first step should be to approximate the curve by a broken (*polygonal*) line. We then try to improve the approximation again and again until the best possible approximation is reached. It is apparent that the main tool we have discussed so far — the limit — will be directly applicable to this improvement process.

We could think of many ways to approximate a curve by a polygonal line, and at first glance it seems that any polygonal line that falls close to the curve would work fine. But the following example shows that the mere requirement of closeness is not enough.

Consider the right triangle of Figure 1.3. The length of the hypotenuse c

is given by the Pythagorean theorem

$$c^2 = a^2 + b^2,$$

where a and b are the perpendicular legs. We take the hypotenuse as the curve that we wish to approximate, and form a "sawtooth" polygonal curve, as shown in the figure. The sawteeth are formed in such a way that each is similar to the original right triangle. We see that the sum of all the legs of the sawteeth equals $a + b$, independent of the number of teeth used. As the number of teeth tends to infinity, the toothed edge appears to come into coincidence with the hypotenuse of the original triangle. This gives the erroneous result $c = a + b$.

So the polygonal line must hold more tightly to the curve somehow. A better idea might be to place all of the polygonal *nodes* (i.e., corners) on the curve (and, of course, this works for a straight line). If a_k is the length of the kth polygonal segment, then the total length of the polygonal curve is the sum

$$L_n = a_1 + a_2 + \cdots + a_n = \sum_{k=1}^{n} a_n.$$

This looks just like the nth partial sum of a series $\sum_{k=1}^{\infty} a_n$. It seems that passage to the limit as $n \to \infty$ in L_n could solve our problem of finding the length of the curve. A difficulty lurks, however: when we summed a series, we added more and more terms *while keeping the previously added terms the same each time*. In the present case, if we increase the number of polygonal segments without bound, then the length of each will tend toward zero. This presents us with a rather strange summation of infinitely small quantities. The situation here is clearly different from that of a series, and we must decide how to deal with such sums. (We will ultimately be led to the definite integral, but we are not ready to discuss that yet.) In addition, we are faced with the idea of a curve being composed of infinitely small pieces, and we need to better understand the implications of this. The problem caught the attention of the philosopher Gottfried Leibniz (1646–1716) during his attempts to explain the structure of natural bodies. Leibniz suspected that any object is composed of infinitely small pieces. This suspicion eventually led him to the notions of differential and derivative. Leibniz actually hoped to use this idea to prove the existence of God through the use of mathematics. The attempt was brave, but from our modern viewpoint, mathematics can only demonstrate the consequences of axioms — not the axioms themselves! But it frequently happens that the errors of great men are so great that they are converted into correct assertions many generations later, as was the case with Hooke's corpuscular theory of light.

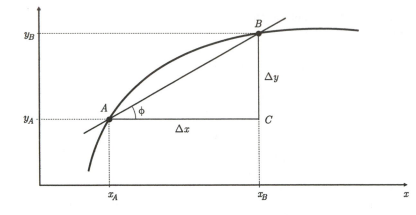

Figure 1.4 Calculation of the tangent line at point A, which is approximated by secant AB constituting an angle ϕ with the positive direction of the x-axis. Note that $\tan\phi = \Delta y/\Delta x$.

Tangent to a Curve

A particular but important problem solved by Leibniz was as follows. Given a curve described by an equation $y = f(x)$, we seek the straight line tangent to the curve at a specified point. Let us first try to reconstruct Leibniz's way of thinking about how such a tangent line should be defined.

On an intuitive basis, it is easy to see what must be done. We must find the straight line that is "closest" to the given curve at the given point. But how should "closeness" be defined in this situation?

- We know what it means for a line to be tangent to a circle: the line must have a unique point in common with the circle. But this definition does not give us a practical way to find the tangent; furthermore, we can easily imagine a curve whose tangent line meets the curve at more than one point. For example, the horizontal line $y = 1$ is tangent to the curve $y = \cos x$ at any $x = 2\pi n$.
- There is a theorem stating that any tangent to a circle is perpendicular to the radius of the circle at the point of tangency. We might be tempted to use this property as a general definition of the tangent line. But where could we take this right angle at a point of an arbitrary curve?
- The tangent to a circle can be imagined as falling at the limit position of a secant line through the desired point of tangency, when the other end of the secant tends to the desired point of tangency. Could this same idea work for a general curve?

Leibniz made the third idea the centerpiece for his definition of the tangent to an arbitrary curve. Following his lead, let us construct a tangent at point A of the curve $y = f(x)$, shown in Figure 1.4. Consider a secant line AB intersecting the curve at point B. This secant and the straight lines

AC and BC running parallel to the coordinate axes determine a triangle ACB. We denote the legs of this triangle by Δx and Δy, as shown, where $\Delta x = x_B - x_A$ is called the *increment* of the argument, and $\Delta y = y_B - y_A$ is the corresponding increment of the function. It is seen that

$$x_B = x_A + \Delta x, \qquad \Delta y = y_B - y_A = f(x_A + \Delta x) - f(x_A),$$

and that the fraction $\Delta y / \Delta x$ is equal to the tangent of the angle whose vertex is A.

Now suppose that B moves down the curve and approaches coincidence with A. We imagine the secant line rotating about A and approaching a limit state. To take this idea further, we must have a function on which we can actually perform the corresponding limit passage. For this, it makes sense to choose the tangent of the angle that the secant makes with the x-direction: this is well defined at each point B that is close to A (but distinct from A), and its limiting value is what we are seeking. Let ϕ_0 be the angle of the tangent at point A of the curve. We have

$$\tan \phi_0 = \lim_{\Delta x \to 0} \frac{f(x_A + \Delta x) - f(x_A)}{\Delta x} = \lim_{\Delta x \to 0} \frac{\Delta y}{\Delta x}. \qquad (1.2)$$

If this limit exists, then we can determine the desired tangent line.

Later, we shall come to understand the importance of (1.2). But first, we will see how the same formula arises in another problem.

Velocity of Motion

Sir Isaac Newton was a scientific giant, an expert in mathematics, physics, economics, and even alchemy. (It is possible that Newton spent more time on the latter than on the sciences for which he is considered a pioneer. His notes indicate a persistent search for the *philosopher's stone* — a stone that would allow one to change various other materials into gold.) Newton tackled the significant problem of finding the velocity of a point in nonuniform motion. We know that velocity in uniform motion is obtained by dividing the distance traveled by the time it takes to travel that distance. But for nonuniform motion, this gives only an *average* velocity. The average is not, for example, the value we see on a car speedometer as we are driving. There were no cars in Newton's time, but there were ships; people were also interested in calculating the motions of the stars and planets. Newton came up with an idea similar to that of Leibniz, as discussed above: instead of trying to calculate the velocity in one step, we seek to approximate it more and more closely via a limiting process. So we shall consider the distance traveled by a point over shorter and shorter time intervals. Each time, we calculate the average velocity as the ratio of distance traveled to time of travel. The limiting value of these average velocities, as the time interval tends to zero, is the *instantaneous velocity* we seek.

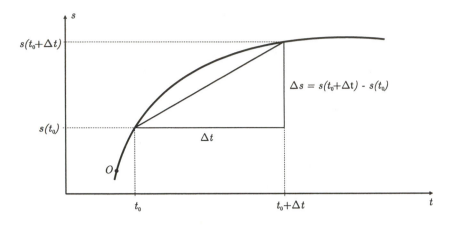

Figure 1.5 Motion of a point along a curve.

Let us realize this scheme. Suppose a point moves along a curve. At time t, it has traveled a distance s along the curve, as measured from the initial point O; hence the motion of the point is given by a function $s(t)$. We find the velocity at a time instant t_0. At this instant, the point is located a distance $s(t_0)$ from O. Now, let a small time interval Δt pass; after this, the point is located a distance $s(t_0 + \Delta t)$ from O. The distance Δs covered during Δt is therefore $s(t_0 + \Delta t) - s(t_0)$, and the average velocity during that time equals $\Delta s/\Delta t$. Figure 1.5 depicts the situation in terms of the new notation we have introduced. Note the similarity to Figure 1.4, and that what we are proposing to call the velocity at the time instant t_0,

$$v = \lim_{\Delta t \to 0} \frac{\Delta s}{\Delta t}, \tag{1.3}$$

is a limiting ratio of the same form as (1.2).

Thus (1.2) and (1.3) coincide in both form and meaning: each reflects the rate of change of some curve. This characteristic of a curve is important, and the construction

$$y'(x) = \lim_{\Delta x \to 0} \frac{y(x + \Delta x) - y(x)}{\Delta x} = \lim_{\Delta x \to 0} \frac{\Delta y}{\Delta x} \tag{1.4}$$

is called the *derivative* of the function $y(x)$ at the point x. Our notation is essentially due to Newton. Leibniz wrote the same thing as

$$\frac{dy}{dx} = \lim_{\Delta x \to 0} \frac{y(x + \Delta x) - y(x)}{\Delta x} = \lim_{\Delta x \to 0} \frac{\Delta y}{\Delta x}. \tag{1.5}$$

Here, he denoted Δx by dx, called it the *differential* of x, and considered it to be infinitely small. The symbol dy stood for the differential of the function, which was also taken to be infinitely small. Note that, in Leibniz's time, there was no theory of limits as we know it today; the related notions were quite fuzzy, and many results were considered to be justified if they gave rise to the same formulas found by other methods (usually more elementary, but not necessarily easier to use; as we know, many problems from arithmetic are easier solved through the use of algebra).

Today, dy is introduced as the simple product

$$dy = y'(x)\,dx,$$

where dx is a simple two-letter notation for the increment of x. Smallness comes only during the limit passage (1.4).

Seeing this important construction involving the limit of a fraction in which the numerator and denominator tend to zero, the reader can understand why the properties of limits are studied so extensively in elementary courses. We shall not discuss the standard formulas for elementary functions, assuming that the reader is at least aware of them.

Let us return to the formula for the derivative. This can be rewritten as

$$\Delta y = y'(x)\Delta x + \omega(|\Delta x|),$$

or

$$y(x + \Delta x) = y(x) + y'(x)\Delta x + \omega(\Delta x),$$

where $\omega(\Delta x)/\Delta x$ tends to zero as $\Delta x \to 0$. This is denoted by a special symbol:

$$\omega(\Delta x) = o(|\Delta x|).$$

In words, we say that $\omega(\Delta x)$ is infinitely small in comparison with $|\Delta x|$ when $|\Delta x|$ tends to zero. So,

$$y(x + \Delta x) = y(x) + y'(x)\Delta x + o(|\Delta x|). \tag{1.6}$$

This shows that the derivative allows us to approximate the behavior of a function in some small neighborhood of a point with a linear function whose behavior is clear. Redenoting x by x_0 and $x + \Delta x$ by x, we can rewrite (1.6) as

$$y(x) = y(x_0) + y'(x_0)(x - x_0) + o(|x - x_0|),$$

and thus the straight line

$$y = y(x_0) + y'(x_0)(x - x_0) \tag{1.7}$$

is a good approximation to the graph of $f(x)$ in some neighborhood of x_0. The last equation means that a function having a derivative at a point behaves close to a linear function in some neighborhood of that point. (See

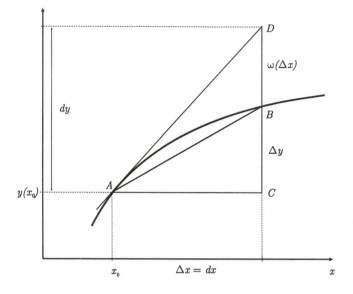

Figure 1.6 Approximating the length of a short segment of a plane curve. Line
AD is described by the equation $y = y(x_0) + y'(x_0)(x - x_0)$.

Figure 1.6.) Our original intention was to find the length of a plane curve
defined by an equation $y = f(x)$. The first step will be to approximate the
length of a small portion of the curve. It appears that the tools we have
introduced are enough for this particular problem. Indeed, let us consider
Figure 1.6, where we show $\Delta x = dx$, Δy, and dy.

The closer Δx is to zero, the better the secant AB fits the portion of the
curve AB. The line AD is tangent to the curve at A. There are two possible
approximations to the length of the curve AB. One is the length of the
hypotenuse AB of triangle ABC, which is

$$\sqrt{\Delta x^2 + \Delta y^2}.$$

A less attractive looking alternative is the length of the hypotenuse of ADC,
which is

$$\sqrt{\Delta x^2 + dy^2} = \sqrt{dx^2 + (y'(x)\, dx)^2} = \sqrt{1 + y'^2(x)}\, dx.$$

For finite Δx, these look quite different; however, if

$$\lim_{\Delta x \to 0} \frac{\omega(\Delta x)}{\Delta x} = 0,$$

as required for the derivative of y to exist at x, then the relative difference between the two approximation lengths disappears in the limit:

$$\lim_{\Delta x \to 0} \frac{\Delta y - y'(x)\Delta x}{\Delta x} = 0.$$

This means that we can use the second approach just as well, and for this we will introduce the length notation

$$ds = \sqrt{1 + y'^2(x)}\, dx. \tag{1.8}$$

We would like the reader to note that any function having a derivative exhibits behavior close to that of a linear function in some neighborhood of any point. This statement extends to functions in many variables, and it long formed the background for physics and mathematical physics: the equations used to describe natural objects and processes were mostly linear. Their derivation involved, as a first step, the idea of linearization based on the change of increments by the differentials. Nonlinear equations, which can model phenomena much more realistically, are difficult to solve, so much of what we know about the vibrations of bodies, and so forth, comes from a linear approach based on the simple considerations above.

Let us turn to the problem of velocity once more. Perhaps the reader has noticed how the speedometer of a car fails to keep up with changes in actual speed. During a quick start, for instance, the speed shown by the speedometer may fall short of the actual speed. The problem is that the speedometer does not measure the car's velocity directly: it cannot perform a derivative operation, as we did above. Instead, it indicates another quantity that relates to velocity through a special type of equation known as a *differential equation*. Such an equation will contain the derivative of a function, and its solution will inherit some of the inertia associated with the actual physical system. When the car moves more or less uniformly, however, the speedometer provides a fairly accurate reading.

Newton (like any physics or math teacher) insisted that problem solving is a useful activity. So let us apply the derivative to an area for which, at first glance, it seems inapplicable: the realm of discrete calculations. Consider the following problem.

A reporter for a local newspaper has published an article entitled "In 200 Years Our Town Will Be One of the Biggest in the World." It cites statistics showing that during each of the last two years the town population grew by 50,000. A simple calculation shows that citizens have been added at a rate of two per second, and 200 years later, there should be an additional 10,000,000.

This reporter did not even realize that he used several important mathematical tools in the process of coming up with this speculation. He first assumed that the law of population increase would follow the present trend. This is a natural assumption, and would be adopted by almost any researcher who is studying something for the first time: what we see is taken to be representative. Then he introduced, at least in the back of his mind, a population growth function something like

$$N = N_0 + 50000\,t. \tag{1.9}$$

Here, N is the number of citizens at time t, where t is measured in years from the time at which the number is N_0. If we take the reporter's statistics to be absolutely correct (which they probably are not), then this equation is valid only for three values of t: $t = 0, 1, 2$. But the reporter assumed it to hold at any time t. The assumption that it holds for any t in the segment $[0, 2]$ is an example of *linear interpolation* of the data; for t outside of $[0, 2]$, we speak of *linear extrapolation*. But extrapolation can get us into trouble. Suppose we use the reporter's linear relation to extrapolate backwards in time (t negative). We might then falsely conclude that a 200-year-old city is only 10 years old (by assuming that 50,000 persons have been added each year to give the present population of 500,000).

Despite his sketchy understanding of population dynamics and his tendency toward empty speculation, the reporter did manage some serious achievements (in addition to increasing his subscription rate): (1) he introduced an approximation of discrete data by a continuous function, and (2) he applied this function outside the region for which his data were obtained. Both of these actions are typical of any science involving discrete processes: economics, biology, continuum mechanics. The assumption of a smooth functional relationship opens the door to many interesting calculations. In particular, the derivative can be found; in the above example, it turns out to be equal to 50,000 (expressed in units of persons per year). This derivative approximates the population growth rate, and can be roughly found by dividing the population increment over a short time (say, a day or a month) by that time interval. Once the derivative is known at a certain time, we can calculate the population at nearby times by using the linear approximation formula (1.7).

But population growth is seldom uniform. The same can be said about the motion of objects; so, to characterize changes in velocity, we introduce the concept of acceleration. This plays the same role with respect to velocity as velocity plays with respect to distance. The acceleration is the derivative of the velocity with respect to time:

$$a = v'(t) = \frac{dv}{dt}.$$

Because $v(t)$ is the derivative of the distance $s(t)$, there must be a relation between $s(t)$ and a: this is denoted by

$$a = s''(t).$$

In this way, we introduce the second derivative of a function $y = f(x)$.

Equivalent notations for this are

$$f''(x), \qquad y''(x), \qquad y'', \qquad \frac{d^2y}{dx^2}.$$

Higher-order derivatives of a function can be introduced similarly.

In elementary physics, we learn that uniformly accelerated motion with acceleration a is described by the equation

$$s = s_0 + v_0 t + \frac{at^2}{2}.$$

We see that the velocity of this motion is

$$v = \left(s_0 + v_0 t + \frac{at^2}{2}\right)' = v_0 + at.$$

The calculation

$$a = v'(t) = \left(s_0 + v_0 t + \frac{at^2}{2}\right)'' = a$$

demonstrates that our present notion of acceleration agrees with that introduced in physics.

Newton's famous law,

$$F = ma = ms''(t),$$

connects the forces acting on an object with the second derivative of the object's position. It took many years for people to understand why the acceleration should appear in this equation instead of the velocity. We might also remark that an exact definition of force cannot be given; the term remains undefined just as with the primary geometric notions of point, line, and plane.

The derivative can be used to investigate the rate of change of any quantity. It also plays an important role in the interpolation and extrapolation of results that are not linear. Let us return to our reporter and suppose that he, finding something wrong with his figures, decides to formulate a more reasonable relation to describe the population growth of his town. He might at first suspect that the greater the population, the greater the growth. So he might take the population increment ΔN to be proportional to the present population $N = N(t)$. He might then suppose that, during a short time Δt, the increment ΔN is proportional to Δt as well. Denoting the proportionality factor by k (which can be calculated for any particular case in which the increments are known), he obtains the relation on $[t, t + \Delta t]$:

$$\Delta N(t) = kN(t)\Delta t. \qquad (1.10)$$

Of course, when he considers different data, he can find that k also depends on N and Δt. But when the dependence is seen to be weak, he reasonably assumes that k is constant. Starting with the initial population $N(0)$ at time $t = 0$, he is now able to find from the equation that the increment $\Delta N(0) = kN(0)\Delta t$ for time Δt, and thus N becomes $N(0) + \Delta N(0)$. Applying the same equation to the time interval $[\Delta t, 2\Delta t]$ with the new "initial" value $N(0) + \Delta N(0)$, he obtains a new increment $\Delta N(\Delta t) = k(N(0) + \Delta N(0))\Delta t$. He can repeat this, and thereby extrapolate N into the future. His results will be more reasonable than those found by linear extrapolation, as long as k remains practically constant (which, unfortunately, will not be the case for long). The reader may have noticed the similarity between this approach and the calculation of compound interest in banking. From a mathematical viewpoint, the two problems are identical.

Equations of the type (1.10) are called *difference equations*. They play an important role in discrete mathematics. But it is also possible to treat continuous processes. What happens if we consider the same equation (1.10) with varying increments Δt? The symbol "Δ" encourages us to apply our knowledge of increments from calculus. So we divide both sides of (1.10) by Δt and find the left-hand side in the form of a difference quotient:

$$\frac{\Delta N}{\Delta t} = kN.$$

Passage to the limit gives us

$$\frac{dN}{dt} = kN.$$

This equation, containing both the unknown $N = N(t)$ and its time derivative, is an example of an *ordinary differential equation*. Although such equations are often difficult to solve, this one is not. We can easily verify that

$$N = ce^{kt}$$

is a solution for any arbitrary constant c. Because we must have $N = N_0$ at the instant $t = 0$, we see that the desired solution is

$$N = N_0 e^{kt}.$$

Taken together with an initial condition such as that above, a differential equation constitutes a *Cauchy problem*.

We believe that the reporter's new assumptions regarding population growth are more reasonable. However, examination of the numbers would predict growth far in excess of his original prediction; in fact, a truly exponential growth rate would mean that sooner or later everyone living on the Earth would have to live in that same town. So the new formula, although more accurate than the old, has a restricted range of applicability (i.e., to relatively short time intervals). Indeed, simplistic assumptions such as strict proportionality can often get us into trouble when modeling an object or process.

1.8 Antiderivatives

Mathematicians always try to create perfect theories. They have their own idea about what is perfect. Particularly valuable are theorems that have an "if and only if" form, because these allow us to reason in two directions. The construction of inverse operations then becomes possible: we can subtract as well as add, divide as well as multiply. It turns out that *differentiation*, the operation of finding a derivative, also has an inverse operation.

Two functions $f(x)$ and $F(x)$ related by the equation

$$F'(x) = f(x) \tag{1.11}$$

are naturally paired in a certain sense. We have already posed the problem of differentiating a given function $F(x)$. The corresponding inverse problem can now be formulated: given $f(x)$, find a function $F(x)$ for which (1.11) holds. Because $f(x)$ is the derivative of $F(x)$, we call $F(x)$ an *antiderivative* (or *integral*) of $f(x)$.

Returning to the mechanics problems we have considered, we see that finding an antiderivative corresponds to finding the distance function $s(t)$ when the velocity function $v(t)$ is given, or finding the velocity when the acceleration is given. In geometrical terms, the process amounts to finding the equation of a curve $y = F(x)$ when the slope $\tan\phi = f(x)$ is given at each point along the curve. It is clear that we can use slope information to deduce the "shape" of a curve, but not its vertical position: we say that we can determine the curve up to a shift along the y-axis (and therefore we can appoint any initial value $y = y_0$ corresponding to $x = x_0$). Hence if $F(x)$ is an antiderivative for $f(x)$, then so is $F(x) + C$ for any constant C. This can be seen by direct differentiation as well. So differentiation does not have a unique inverse operation. However, it turns out that this issue of an additive constant is the only one we need to worry about: if $F_1(x)$ and $F_2(x)$ are any two antiderivatives for $f(x)$, then there is a constant C such that $F_2(x) = F_1(x) + C$.

The class of all antiderivatives for a function $f(x)$ is called an *indefinite integral* of $f(x)$. It is denoted in a special way:

$$\int f(x)\, dx.$$

The integral symbol $\int$ was inherited from the letter "S" in the word "Sum" (although we have yet to explain what sum we are referring to). The symbol

dx is actually the differential of x, which plays a crucial role when one decides to change variables in the integral.

Calculus students become familiar with extensive tables of the derivatives of elementary functions. They also learn simple rules for differentiating functions that can be constructed by combining the elementary functions in various ways (e.g., through the use of arithmetic operations and composition). It has been said that one could teach a monkey to differentiate in this way. This is not to imply that all integration problems are simple. If we have two functions $f_1(x), f_2(x)$ for which the integrals are known, we can easily integrate their linear combination:

$$\int \left(c_1 f_1(x) + c_2 f_2(x) \right) dx = c_1 \int f_1(x) + c_2 \int f_2(x) \, dx.$$

But no such formula exists for

$$\int f_1(x) \, f_2(x) \, dx \qquad \text{or} \qquad \int \frac{f_1(x)}{f_2(x)} \, dx.$$

Nor is there a general formula for handling composite functions of the form $f(g(x))$ — that is, when f depends on g, which is a function $g = g(x)$ itself. Because of this, the list of known antiderivatives is short, the extensive appearance of some integral tables used by mathematicians and engineers notwithstanding. Many nice-looking functions cannot be integrated analytically (*in closed form*). But some functions turn out to be so important that we must learn to approximate their integrals through the use of computer methods. These include the so-called *special functions* of mathematical physics, the properties of which have led to remarkable insight about natural phenomena.

We shall not calculate long lists of specific integrals, but we should mention that most textbooks make extensive use of two main formulas. One is the formula for a change of variable:

$$\int f(g(x)) \, g'(x) \, dx = \int f(u) \, du,$$

where $u = g(x)$. The other is the formula for *integration by parts*:

$$\int g'(x) f(x) \, dx = g(x) f(x) - \int g(x) f'(x) \, dx,$$

which follows from the product rule for differentiation. Firm statements can also be made regarding the *existence* of derivatives and integrals (although we may be unable to represent these in terms of elementary functions). For example, every continuous function has a continuous antiderivative. But not all continuous functions can be differentiated everywhere: $y = |x|$ is not differentiable at $x = 0$, for example. In the next section, we shall see how

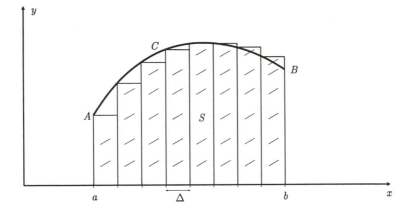

Figure 1.7 Approximation of area using narrow rectangles. The area sought lies between curve ACB and the x-axis.

an antiderivative can be constructed for a continuous function, and even for a *piecewise continuous* function whose graph displays a finite number of jumps. In both of these cases, the antiderivative turns out to be continuous. Roughly speaking, differentiation can worsen the smoothness properties of a function, while integration generally improves them.

1.9 Definite Integral

We are still interested in finding the length of a curve given by an equation. Consideration of this problem is leading us through the main points of calculus. One important topic remains: the definite integral. All major rivers start from small springs, and the same can be said of the great mathematical theories. So let us consider a problem that gave rise to the notion of the definite integral. We seek the area of the curvilinear trapezoid shown in Figure 1.7. This problem yields to our earlier approach: approximation, limit passage, justification.

How should we approximate the area of the trapezoid? It is clear that we should make use of some elementary shapes having known areas. Here, the simplest and most appropriate shape is the rectangle, for which the area equals the product of the width Δ and height h: $S = h\Delta$. We shall use narrow rectangles in order to fit the trapezoid closely. So we partition the segment $[a, b]$ into n parts each of length $\Delta = (b-a)/n$, and draw rectangles with bottoms resting on the partition intervals. The height of each rectangle is taken to be equal to the value of the function $f(x)$ at the left endpoint of the base segment, as shown in Figure 1.7.

We see that some rectangles fall completely within the trapezoid, while others do not. The smaller we make Δ, the less relative difference we will have between the area of the trapezoid and the total area of all the approximating rectangles. Denoting the height of the ith rectangle by $y_i = f(a + (i-1)\Delta)$, we get an expression for the area of approximation:

$$S_n = y_1\Delta + y_2\Delta + \cdots + y_n\Delta = \sum_{i=1}^{n} y_i\Delta.$$

Because we now have a sequence $\{S_n\}$, we can attempt a limit passage as $n \to \infty$. This gives some limit value

$$S = \lim_{n\to\infty} S_n$$

that we may consider as a candidate for the area of the curvilinear trapezoid. In fact, it may seem obvious that S *is* the area sought, but we shall be careful and try to understand what happens if we take another approximation. For example, we could redefine the heights of the elementary rectangles as the values of $f(x)$ at the *right* endpoints of the base segments: $y_i = f(a + i\Delta)$. This would give a new expression for S_n, and we could perform another limit passage. Suppose we thereby obtained the same value of S: should we be convinced that the area sought is this value? The answer is that we should be *encouraged*, but not convinced. We should really try all possible variants. We could, for example, take the rectangle heights to be the values of $f(x)$ at intermediate points of the basal segments. For each choice, we get a corresponding sum S_n; if each of these tends to the same limit S as $n \to \infty$, then we can be even more convinced that S is the area of the curvilinear trapezoid. Indeed, among those approximating sets of rectangles, there are some that fully contain the trapezoid and others that are fully contained within it; by our understanding of area we expect the area of the trapezoid to fall between the areas of two such approximating sets. If the corresponding limit passages close in on the same limit S, then probably everyone except a mathematician would be convinced that S is the area sought. The latter might object to the fact that we took only uniform partitions of $[a, b]$: "I can show you a function for which S stays the same for all uniform partitions but becomes quite another value for nonuniform partitions. Do you wish to see it?" We would be wise to decline as the example is not nice. However, what should we do? Since the time of Newton and Leibnitz, all but a few persons have been satisfied with the level of rigor of this theory in each stage of its development. It turns out that this little struggle took almost 200 years to resolve. The end result was the notion of the *Riemann integral*, due to Georg Riemann (1826–1866).

Definition 1.5 The definite integral of a function $f(x)$ over the segment $[a, b]$ is the (common) limit of all "Riemann sums":

$$S = \lim_{n\to\infty} \sum_{i=1}^{n} f(t_i)\Delta_i.$$

Here the x_i are the points of a partition of $[a, b]$,

$$a = x_1 < x_2 < \cdots < x_{n+1} = b, \qquad \Delta_i = x_{i+1} - x_i,$$

and each t_i is an arbitrary point of the subinterval $[x_i, x_{i+1}]$. For the nth partition, we denote by δ_n the length of the longest subinterval: $\delta_n = \max_i \Delta_i$. We stress that the limits must be taken for arbitrary partitioning of $[a, b]$ and arbitrary choice of the points t_i such that $\delta_n \to 0$ as $n \to \infty$. This limit value, the definite integral, is denoted by the symbol

$$\int_a^b f(x)\, dx.$$

Note that in Figure 1.7 we depicted a positive function $f(x)$. In this case, the definite integral yields the area of the enclosed curvilinear trapezoid. It is clear that if $f(x) < 0$, then the integral will yield a negative value for S.

The way in which we introduced rectangle approximations of the trapezoid tells us that if we break $[a, b]$ into two parts with an intermediate point c, then we can similarly calculate two definite integrals,

$$\int_a^c f(x)\, dx \qquad \text{and} \qquad \int_c^b f(x)\, dx,$$

for which the sum of the approximating finite sums gives an approximation to the integral over $[a, b]$. Thus,

$$\int_a^b f(x)\, dx = \int_a^c f(x)\, dx + \int_c^b f(x)\, dx. \tag{1.12}$$

Using this, we conclude that if $f(x)$ is nonnegative on $[c, b]$ and nonpositive on $[a, c]$, then the area "between" the curve $y = f(x)$ and the x-axis is given by the formula

$$S = -\int_a^c f(x)\, dx + \int_c^b f(x)\, dx.$$

It is clear that the property of linearity we encountered for the indefinite integral is valid as well:

$$\int_a^b (c_1 f_1(x) + c_2 f_2(x))\, dx = c_1 \int_a^b f_1(x) + c_2 \int_a^b f_2(x)\, dx.$$

This is not a simple coincidence; the similarity between the definite and indefinite integrals is not just in name. But before pursuing this, we should mention that the notion of the definite integral, its properties, and the main formulas for its calculation were found by Newton and Leibniz almost simultaneously. Both were aware of the complexity that would be involved in full validation of these tools, but they acted as do all pioneers in new areas: they used the solution of one problem to conquer the next, satisfied in the fact that their new methods gave results that could in certain cases be verified by comparison with results obtained by older methods. But the powerful intuition of these men assured them that everything was correct. (Of course, the history of mathematics contains many other examples in which authors were sure of their correctness on certain conjectures that were later disproved.) Full validation came much later, but early mathematicians nevertheless proceeded to solve many important problems using the definite integral. Newton and Leibniz each claimed credit for its development; their argument went on for decades and was later continued by their friends and students. But it appears that they discovered everything independently. Interestingly, a substantial claim on the discovery of the definite integral could be made by the ancient Greeks, for the works of Archimedes (c. 287–212 B.C.) contain problems essentially solved by this notion. Later mathematicians managed to solve problems involving areas, lengths, and volumes so complicated that it is unclear how they could have succeeded without the definite integral. So perhaps the definite integral really existed before Newton and Leibniz.[3] However, the notion as presented by Newton and Leibniz really started a kind of revolution. Suddenly even those persons not considered geniuses were able to solve complex problems using the new tools that had been exposed.

What we have said above sounds very nice: we have proposed some limiting process to define an area, but surely cannot expect that an ordinary person could use this description to perform an actual calculation given a formula for $f(x)$. Moreover, according to the definition, we must perform infinitely many limit passages to justify a result. Fortunately, it can be shown that the Riemann integral exists for any continuous or piecewise continuous function. This means that it suffices to try only one limit passage in calculating a result. The next piece of good news is that there is a formula that connects the definite integral of $f(x)$ with its antiderivative $F(x)$:

$$\int_a^b f(x)\,dx = F(b) - F(a).$$

This is the *Newton–Leibniz formula*. For calculations, it is convenient to present the right-hand side in the form

$$F(b) - F(a) = F(x)\Big|_a^b,$$

---

[3]It is not uncommon that as soon as someone gains notoriety for publishing an important result, many people claim that they accomplished the same thing earlier but decided not to publish it because it seemed unimportant.

and thus to write

$$\int_a^b f(x)\, dx = F(x)\Big|_a^b.\qquad(1.13)$$

So a good table of antiderivatives can enable us to calculate many (but by no means all) definite integrals.

The definite integral satisfies two main relations that resemble the formulas we quoted for indefinite integrals. There is a formula for integration by parts:

$$\int_a^b g'(x)f(x)\, dx = g(x)f(x)\Big|_a^b - \int_a^b g(x)f'(x)\, dx.\qquad(1.14)$$

There is also a formula for change of variables:

$$\int_a^b f(g(x))\, g'(x)\, dx = \int_c^d f(u)\, du,\qquad(1.15)$$

where $c = f(a)$ and $d = f(b)$. We shall not give further explanation of these formulas, or rules for their usage.

We should, however, explain why the symbol $\int$ was used for the definite integral just as it was used for the indefinite integral. We can do this most easily by allowing the upper limit b in the definite integral to become a variable parameter. It is clear that the value of $\int_a^b f(x)\, dx$ depends uniquely on a and b, hence it is a function of these two variables. Fixing a, we get a function of the single variable b, which we redenote by x:

$$F(x) = \int_a^x f(t)\, dt$$

(of course, we had to redenote x in the integrand somehow, and for this we used t). We will show that $F(x)$ is an antiderivative for $f(x)$ at all points of continuity of the latter. For this, we need to demonstrate that $f(x) = F'(x)$. We shall use the definition of the derivative. First, we calculate the increment of $F(x)$ due to the increment Δx of x:

$$\Delta F(x) = F(x + \Delta x) - F(x) = \int_a^{x+\Delta x} f(t)\, dt - \int_a^x f(t)\, dt.$$

By (1.12), we have

$$\int_a^{x+\Delta x} f(t)\, dt = \int_a^x f(t)\, dt + \int_x^{x+\Delta x} f(t)\, dt,$$

and so

$$\Delta F(x) = \int_x^{x+\Delta x} f(t)\, dt.$$

In the limit we should have Δx small. (We may imagine it to be positive for geometrical clarity, but this is not necessary.) Thus, for small Δx, we have the area of a narrow trapezoid (like a partition portion in Figure 1.7), whose area can be approximated by a rectangle with height taken as the value of $f(x)$ at a point of $[x, x + \Delta x]$. In particular, if we take x as this point, we get

$$\Delta F(x) = f(x)\Delta x + o(|\Delta x|).$$

The difference between $\Delta F(x)$ and its "rectangular" approximation $f(x)\Delta x$ is $o(|\Delta x|)$ by continuity of $f(x)$ at x. Hence,

$$F'(x) = \lim_{\Delta x \to 0} \frac{\Delta F(x)}{\Delta x} = \lim_{\Delta x \to 0} \frac{f(x)\Delta x + o(|\Delta x|)}{\Delta x} = f(x),$$

as desired.

The fact that the definite integral $F(x) = \int_a^x f(t)\,dt$ with variable upper limit is an antiderivative is the basis for proving the Newton–Leibniz formula (1.13). Indeed,

$$F(b) = \int_a^b f(t)\,dt.$$

It is clear that the area of a trapezoid with a zero-length bottom is zero; thus,

$$F(a) = \int_a^a f(t)\,dt = 0.$$

Subtracting $F(a)$ from the above, we have

$$\int_a^b f(t)\,dt = F(b) - F(a).$$

But for any other antiderivative $\Phi(x)$ for $f(x)$, we have

$$\Phi(x) = F(x) + C,$$

and, therefore,

$$\Phi(b) - \Phi(a) = (F(b) + C) - (F(a) + C) = F(b) - F(a),$$

which justifies the Newton–Leibniz formula (1.13) completely.

Thus, starting with a particular problem, we have introduced a powerful tool called the definite integral. We will see it applied to the solution of various problems.

Improper Integrals

The practice of reasoning using graphs can be illustrative but dangerous. On a graph, we normally depict a function with typical but rather nice behavior, and this can trick us into making unwarranted assumptions. Many cases unfortunately fall outside the realm of graphical presentation; hence when we use a result in applications, we sometimes run into difficulties not suggested by the figure. In applied mathematics, we are interested in both the ordinary and *singular* cases, and more modern research has been largely concerned with the latter.

With integration, there are two particular but important cases that cannot be covered by the construction of Figure 1.7. The first is when one or both of the endpoints of the segment $[a, b]$ is infinite. Suppose, for example, that the integration interval is $[a, \infty)$. Note that we "exclude" the infinite endpoint. Despite the fact that we use the symbol ∞ like the symbol of a point and, moreover, sometimes consider it as a point (we can change the variable x to $t = 1/x$, finding under this change that the point $t = 0$ corresponds to the "point" ∞ for x; hence if we regard this transformation as "point-to-point" we can formally refer to ∞ as a point), a point at infinity has many strange properties. For example, a "small" neighborhood of it (which really is small in terms of the transformed variable t) is quite infinite: $x > M$. The integral over $[a, \infty)$ is denoted by

$$\int_a^\infty f(x)\, dx$$

and is said to be *improper*. We cannot draw an infinite graph for $y = f(x)$ and use this to construct the integral. However, we can remove an infinite "tail" $x > N$ from the graph, calculate the integral $\int_a^N f(x)\, dx$, and then let $N \to \infty$:

$$\int_a^\infty f(x)\, dx = \lim_{N \to \infty} \int_a^N f(x)\, dx.$$

If the limit on the right exists, then this formula gives the value of the improper integral and the latter is said to be convergent; if not, the integral is said to be nonconvergent.

The reader can reformulate this definition for the cases in which (1) the left endpoint of the integration interval is infinite, and (2) both endpoints are infinite. Also not covered by Figure 1.7 is the case in which the integration interval contains a point at which the function tends to infinity. We suppose there is only a finite number of such points; this means we can split the integral into several integrals and thereby place each singular point at an endpoint of an integration interval. An example of such a point is $x = 0$ for the function $y = 1/x$. Trying to approximate the area under the curve $y = f(x)$, we can draw "right" rectangles but not "left" rectangles because at $x = 0$ the value of $1/x$ is undefined (we would say it is $+\infty$ in this situation). So, in this case, the infinite "tail" having $y > M$ cannot be covered by the approximating rectangles and we need to circumvent this unpleasant situation. The approach is similar to that described above. Suppose we would like to calculate $\int_a^b f(x)\,dx$, where $f(x)$ is continuous in $[a, b)$ but has no limit when $x \to b$ while x remains in $[a, b)$. Let us remove the part of the figure that lies in the domain $x > b - \varepsilon$, $\varepsilon > 0$, and calculate

$$\int_a^{b-\varepsilon} f(x)\,dx.$$

If this possesses a limit as $\varepsilon \to 0$ ($\varepsilon > 0$), then it is natural to call that limit the improper integral

$$\int_a^b f(x)\,dx.$$

Again, this is said to be convergent if the limit exists, nonconvergent otherwise. The reader can verify that $\int_0^1 1/x\,dx$ is nonconvergent and $\int_0^1 1/\sqrt{x}\,dx$ is convergent. This definition covers not only the case in which $f(x)$ tends to infinity, but that in which the function behaves like $y = \sin(1/x)$ at $x = 0$. In the latter case, we cannot properly approximate the figure with finite rectangles.

Because most antiderivatives cannot be found in analytical form, we often resort to numerical approximations of integrals. Although the use of computer methods is now common, we must understand that these can only implement a finite number of computations. For this reason, the issue of convergence remains important even in "practical" approximate work. Any good calculus textbook will contain various sufficient conditions for convergence or nonconvergence of integrals. Many are based on a principle of comparison: we compare the integrand with some standard function for which we know the behavior of the corresponding integral. But much more important is the question of whether it is possible to calculate a given improper integral to a desired degree of accuracy.

1.10 The Length of a Curve

Now we can return to the problem of finding the length of a curve given by the equation $y = f(x)$. We have proposed an approximation to the length of a small piece of the curve that corresponds to a small segment $[x, x + \Delta x]$: it is

$$\sqrt{1 + f'^2(x)}\,\Delta x.$$

We have also found that this differs from the length of the chord from $(x, f(x))$ to $(x + \Delta x, f(x + \Delta x))$ by a value that is infinitely small in comparison with $|\Delta x|$, so the length of the chord can be written as

$$\left(\sqrt{1 + f'^2(x)} + \omega(x, \Delta x)\right) \Delta x,$$

where $\omega(x, \Delta x) \to 0$ when $\Delta x \to 0$. If the expression $\sqrt{1 + f'^2(x)}$ is a continuous function on $[a, b]$, then it can be shown that $\omega(x, \Delta x)$ tends to zero *uniformly*, that is, in a way that is independent of x. Put another way, there is a function $\eta(\Delta x)$ such that $|\omega(x, \Delta x)| < \eta(\Delta x)$ and $\eta(\Delta x) \to 0$ as $\Delta x \to 0$. This allows us to determine the needed length. We begin with the length of a polygonal curve inscribed into the given curve at nodes equispaced along the x-direction. Let this internode spacing be Δ. Denote $x_i = a + (i - 1)\Delta$ for integers i. Then the length of the polygonal curve is

$$S_n = \left(\sqrt{1 + f'^2(x_1)} + \omega(x_1, \Delta)\right)\Delta + \cdots + \left(\sqrt{1 + f'^2(x_n)} + \omega(x_n, \Delta)\right)\Delta$$

$$= \sum_{i=1}^{n} \sqrt{1 + f'^2(x_i)}\Delta + \sum_{i=1}^{n} \omega(x_i, \Delta)\Delta.$$

Now we introduce the length of the curve as

$$S = \lim_{n \to \infty} S_n.$$

Observe that this is a definition: we cannot "prove" that S is the length. We can only verify its appropriateness for "elementary" curves (straight segments, circular arcs, parts of ellipses, etc.) whose lengths are known through other approaches. Because the length of a curve in mathematics is an exact number, we cannot verify the formula experimentally as is customary in engineering: any measurement involves some uncertainty.

Let us see what happens in the limit passage. Because each $\omega(x_i, \Delta)$ tends to zero uniformly with respect to x, when $n \to \infty$, we get

$$\left| \sum_{i=1}^{n} \omega(x_i, \Delta)\Delta \right| < \eta(\Delta) \sum_{i=1}^{n} \Delta = \eta(\Delta)(b - a) \to 0.$$

So the limit of the sum $\sum_{i=1}^{n} \sqrt{1 + f'^2(x_i)}\Delta$ gives S. But we saw this construction before, and we can write

$$S = \int_a^b \sqrt{1 + f'^2(x)} \, dx. \tag{1.16}$$

Remembering what we said about the definite integral of a continuous function, we can state that the same value S will be obtained in the limit passage of the above type if the nodes are not equispaced along x, or if the values for $\sqrt{1 + f'^2(x)}$ are taken at any intermediate points of the corresponding small segments on the x-axis. For this, it is sufficient to suppose that $\sqrt{1 + f'^2(x)}$ is a piecewise continuous function on $[a, b]$.

We have seen how simple it is to introduce the length of a curve through the use of such a powerful tool as the definite integral. In a similar fashion we could introduce the length of a plane curve given parametrically by

$$x = x(t), \qquad y = y(t).$$

It is

$$S = \int_a^b \sqrt{x'^2(t) + y'^2(t)}\, dt,$$

where $[a, b]$ is the interval over which the parameter t must change to describe the curve. Similarly, the volume of a figure of revolution of the curve $y = f(x)$, given on $[a, b]$, about the x-axis is

$$V = \pi \int_a^b f^2(x)\, dx. \tag{1.17}$$

In the next section, we shall consider the latter more carefully.

It is worth mentioning that our use of uniformity here is an example of a common practice in many textbooks: the conditions under which something is proved are scattered inside the proof, so anyone who wishes to use this result in his or her own work should gather all the conditions together first. Sometimes the conditions are scattered over previous pages as well, so a potential end user must learn the whole book first. This is why the rules of politeness in mathematics require one to formulate explicit statements of theorems, including any and all conditions required for the theorem to hold. But inclusion of all necessary assumptions would overcomplicate the statements of many simple theorems, often requiring enumeration of conditions going all the way back to the axioms of the subject. So a good sense of judgment is required. The bad news is that one cannot make use of theorem statements if one does not know exactly what conditions have been assumed but suppressed, and, as a result, one must learn the theory in total. Fortunately, the purpose of our present discussion is the presentation of only major ideas rather than detailed techniques for application.

1.11 Multidimensional Integrals

Among mathematicians, there is a general viewpoint that everything should be done in a strictly formal manner. This way everything is expressed in formulas, intuitive considerations are not involved, and the reader encounters long chains of implications punctuated by new definitions, theorems, and lemmas. But it is possible that if the pioneers of mathematics held to this practice, then we would not have any mathematics. If we consider the history of any branch of mathematics, we never see purely abstract problems; all of the baseline problems came from the real world, as did the methods for their solution. The calculus of variations originated as a series of optimization problems, graph theory originated via the famous problem of the seven bridges of Köningsberg, and so on. Mathematicians often claim to solve problems appropriate to an ideal world, but these problems have their roots in our ordinary world. The latter may not be so ideal, but it is certainly richer in forms, colors, and textures, and reasoning with models of the real world can help us understand abstract problems and find ways to solve them. Even persons of genius have solved problems using the tools that they discovered along the road of life. Everything can ultimately be traced back to earthly origins.

To discuss how multidimensional integrals appear, let us consider a special real-world problem. Suppose we have a nonhomogeneous cube with edge length a, and that we know the mass density at each point of the cube. Imagine that $a = 1$ km, so that nobody can weigh the cube directly. We need to determine its total mass.

In this problem, we find interesting extensions of the notions we have considered. First, let us understand what is meant by the density of a material. In elementary physics, they say it is the mass of a unit cube: it is therefore written as $\rho = m/V$, where m is the mass of a volume V. If we have a nonhomogeneous body of volume V, then this simple formula gives us the *average* density of the body. But how to define the density *at a point*? We use the same idea we used to define instantaneous velocity. We take a small piece of material containing the point of interest, and calculate the average density for that piece. We then repeat for a smaller piece containing the point. The smaller the volume we take, the closer the average density is to what we need. Thus it makes sense to produce the limit passage and to declare that the density at a point is

$$\rho = \lim_{V \to 0} \frac{m(V)}{V},$$

where V is the volume of the portion of the body containing the point, $m(V)$ is its mass, and the limit should be the same for any sequence of portions of the body whose volumes are V_k and that contracts down to the point as $V_k \to 0$.

We see that this construction is close to that for the derivative, where the roles of the increments are played by $m(V)$ and V. From a physical viewpoint, the limit passage is not nice, because at some stage we will find ourselves dealing with the world of individual atoms. So an abstraction comes in, where we agree to consider a "continuous" medium that has no empty interatomic spaces. Such abstractions are not absolutely precise, but in the natural sciences we always deal with approximations. The notion of density at a point is used in any physical science where nonuniformly distributed materials or fields occur.

Now, when we think of the density as an average density for a small portion of a cube with side length Δ, it is natural to introduce a fine partition into the cube, cutting the cube into small cubes whose volumes are Δ^3. We then calculate the approximate mass of each cube of the partition as $\rho_i \Delta^3$, where ρ_i is the density at some interior point of the small cube, and sum up all of these quantities. Thus, we approximate the real mass as

$$M_n = \sum_{i=1}^{n} \rho_i \Delta^3.$$

The next step is the limit passage as n (the number of small cubes of the partition) tends to infinity. This brings us the value M that we expect to be the mass of the cube. The structure of the expression for M_n is so similar to the nth Riemann sum for the definite integral that we can assume (correctly) we are witnessing some extension into space theory of the notion of the definite integral in one variable.

To produce a spatial analogue, we need to introduce a coordinate frame. We let this be Cartesian and write the volume of an elementary cube as $\Delta x \, \Delta y \, \Delta z$. Remembering that we denote the increment of the independent variable in differential notation, we can write

$$M = \lim_{n \to \infty} M_n = \iiint_V \rho(x, y, z) \, dx \, dy \, dz. \tag{1.18}$$

We have talked about a cube, but the same operation can be done with any "reasonable" three-dimensional figure; instead of a simple partition on each step, merely inscribe an approximating set of small cubes into the initial volume V. To validate such a definition, we should verify all limit passages for which the approximating small cubes become rectangular parallelepipeds and for which the value used for approximating each ρ_i is taken at any point inside the elementary parallelepiped. If, in any case, we have the same result M, then we claim that it is the mass of the figure, and on the right-hand side of (1.18) we have the three-dimensional integral of $\rho(x, y, z)$ over the volume V.

By its physical meaning, ρ must be nonnegative at any point (a value of zero for the density is handy when we wish to write the integral over the whole space, in which case we put $\rho = 0$ at points where there is no material). However, it is clear that in this construction of the three-dimensional integral we can consider nonphysical negative values for ρ, and thus we can introduce the abstract definite integral of a function $f(x, y, z)$,

$$\iiint_V f(x, y, z) \, dx \, dy \, dz,$$

in the same manner as we did for ρ. This exists if V is not too "wild" and $f(x, y, z)$ is piecewise continuous on V. In theory, there are domains that cannot be approximated by a finite number of parallelepipeds, and through further study one can learn conditions sufficient for a domain to be "suitable" for Riemann integration. The good news is that domains with boundaries described by functions that occur in calculus textbooks are all "suitable."

In a similar way, we can construct a definite integral over a two-dimensional domain (not necessarily planar) or even over an n-dimensional domain. These constructions, as well as the theory of improper definite multidimensional integrals and their reduction to iterated one-dimensional integrals, can be found in standard calculus textbooks.

In the next section, we shall consider some practical aspects of the calculation of definite integrals.

1.12 Approximate Integration

Constructing the notion of the definite integral, we began with a reasonable idea of how certain quantities should be approximated. Over a long period in the history of mathematics, many attempts were made to calculate definite integrals using their approximating property; however, the required effort was too large for even simple integrals. So the efforts of mathematicians were directed toward obtaining analytical formulas. But history is a snake that likes to catch its own tail: with the advent of the computer, the approximation of definite integrals is now commonplace; on the other hand, analytical calculations of integrals are produced more by computer programs than with pencil and paper.

We will not describe all the nice aspects of performing integration on a computer. (As is so often done, we will say in effect, "Things are good but ...," and then proceed to elaborate on how bad things are!) Computers are fast but can do only finite numbers of calculations involving only finite numbers of digits. The same can be said of humans; but computers are truly dumb, and if the user does not provide the machine with foolproof instructions, then it will do a poor job. So a computer user must know what he or she is doing. Now let us discuss some consequences of all this for the calculation of definite integrals.

The first step for defining the definite integral $\int_a^b f(x)\,dx$ was construction of the finite sum

$$S_n = \sum_{i=1}^{n} f(a + (i-1)\Delta)\,\Delta, \qquad \Delta = \frac{b-a}{n}. \tag{1.19}$$

Then we stated that the limit passage as $n \to \infty$ yields the integral for a continuous function $f(x)$. This means that the more terms we take in the sum the closer it is to the needed value. At first glance it seems we must increase n as much as possible if we need high accuracy. This turns out to be a bad idea: a computer cannot sum an infinite number of terms. First, every computer has a smallest number that it can represent: this is known as its *machine zero* or *machine ε*, a number often small but nonetheless finite.[4] Thus Δ, the step-size of integration, can be no smaller than this machine ε. There is more bad news: during calculations, a computer chops off all decimal places beginning at a certain position. In the process of adding sufficiently many terms to the sum, a systematic chopping error can produce a large total error for a large number of steps. Hence the approximate calculation of an integral (and this holds not only for integration but for any calculation that involves a great number of actions with chopping of intermediate results) has a property of "step optimality": there is a certain N such that when $n < N$, the sums S_n approach the real value more and more closely, but when $n > N$, the error begins to increase and can reach any level. Thus, for any computer, there is some number N by which the number of elementary operations is restricted. To make matters worse, the calculation of integrals is often just a small part of a much larger calculation problem. In such cases, it may be necessary to calculate thousands of integrals.

All of this means that a simple, straightforward calculation of a definite integral by means of the formula (1.19) is not the best way. We need not only more accurate schemes, but more efficient ones. The unfortunate persons who once had to calculate integrals using only pen and paper had a much better grasp of this than do most present-day mathematics students.

It turns out to be rather challenging to develop a scheme with better accuracy. We could begin by simply trying several different types of approximation. We would find that the right rectangle approximation

$$S_n = \sum_{i=1}^{n} f(a + i\Delta)\Delta$$

increases our integration accuracy for some functions and decreases it for others. One idea is to take the average of the right and left rectangle approximations: we add the approximation above to (1.19) and divide the result by two. This gives a useful formula known as the *trapezoidal rule*. Here, the area bounded by a curve is approximated by means of the area under a polygonal line with nodes on the curve $y = f(x)$, which can be viewed as the total area enclosed in a series of trapezoids.

We could go on attempting to improve the approximation, just as was done throughout the long history of numerical analysis. But let us consider the general principles under which one may construct an economical and accurate scheme to approximate the definite integral. After all, an understanding

---

[4]Certain practitioners like to use the term "infinitely small." Many mechanical engineers would probably consider anything smaller than 10^{-6} m to be such. However, no finite number can be infinitely small or infinitely large. In fact, a finite number will be automatically large in some circumstances and small in others. A 0.1 mm growth of someone's fingernail could be thought of as infinitely small, but the same length is rightly regarded as almost infinite when discussing certain aspects of the atomic nucleus.

of general principles frees us from having to remember details.

Although the definite integral is a relatively simple notion, there are different principles by which the optimality of its approximation is judged. We begin with the local principle of approximating an elementary curvilinear trapezoid over a segment $[a, b]$. We do not require this segment to be small; we need a formula for approximating a trapezoidal area that uses few intermediate values of $f(x)$ on $[a, b]$, according to some criterion of optimality.

The criterion can be as follows: formulas for approximating the definite integral of $f(x)$ over $[a, b]$ must be exact for all polynomials of some fixed degree. This was not the only criterion proposed for deriving approximation formulas. But we shall restrict ourselves to it.

The reader can verify that the formula $\frac{1}{2}(f(a) + f(b))(b - a)$ gives the correct value for the definite integral of any linear function $y = kx + c$. However, it fails for quadratic polynomials such as $y = x^2$.

We know that polynomials of sufficiently high order can approximate a continuous function to within any accuracy on any segment. If the segment is sufficiently small, an accurate approximation of a function is possible with a polynomial of low order.

This means that on a small segment we can approximate the integrand function by a parabola having the form $y = a_0 x^2 + a_1 x + a_2$. The integral of this function is known exactly. If we have three points (x_i, y_i) with different x_i, then we can draw the unique parabola through them, substituting the coordinates of the points into the parabola equation and thereby obtaining three independent equations in the three unknowns a_i. It is sensible to use for approximation the endpoints of the graph of $f(x)$. Because there is no reason why one endpoint of $[a, b]$ should be preferable to the other, as the third point we choose the middle point $(a + b)/2$. Because the value of the integral depends on $f(x)$ linearly, we expect that the dependence of the approximating formula on the values of $f(x)$ at the points a, b, and $(a+b)/2$ should be linear, so the formula has the form

$$\int_a^b f(x)\, dx \approx d_1 f(a) + d_2 f(b) + d_3 f((a + b)/2).$$

This equation, by the above principle, must be exact when we take $y = a_0 x^2 + a_1 x + a_2$ with arbitrary a_i. Substituting this into the equation and equating the coefficients of the a_i, we have three linear equations

$$b - a = d_1 + d_2 + d_3,$$

$$\frac{b^2}{2} - \frac{a^2}{2} = d_1 a + d_2 b + d_3 \frac{a + b}{2},$$

$$\frac{b^3}{3} - \frac{a^3}{3} = d_1 a^2 + d_2 b^2 + d_3 \left(\frac{a + b}{2}\right)^2,$$

which can be easily solved to yield

$$\int_a^b f(x)\,dx \approx \frac{b-a}{6}\left(f(a)+4f((a+b)/2)+f(b)\right).$$

This formula is used on a small segment $[a,b]$ and gives a better approximation than that afforded by a linear function. Now note that the coefficients of the approximation depend on the difference $b-a$ only, and this allows us to construct a formula for a general segment $[a,b]$ by subdividing it into $2n$ equal portions of length $h=(b-a)/(2n)$. The result is *Simpson's rule*:

$$\int_a^b f(x)\,dx$$
$$\approx \frac{h}{3}\left[y_0+y_{2n}+4\left(y_1+y_3+\cdots+y_{2n-1}\right)+2\left(y_2+y_4+\cdots+y_{2n-2}\right)\right],$$

where $y_i=f(a+ih)$, $i=0,1,\ldots,2n$. This has a better degree of accuracy than the trapezoidal rule. The accuracy estimates we find in textbooks for Simpson's rule are based on the fact that over each partition segment $[x_i,x_i+2h]$ we can use Taylor's formula for the function $f(x)$:

$$f(x)=f(x_i)+f'(x_i)(x-x_i)+\frac{1}{2}f''(x_i)(x-x_i)^2+\frac{1}{6}f'''(\xi)(x-x_i)^3,$$

where ξ lies in the segment $[x_i,x_i+2h]$. Here, we use Lagrange's form of the remainder, and the formula is exact when $f'''(x)$ is continuous on the segment. Because the inaccuracy of the three-term approximation

$$f_h(x)=f(x_i)+f'(x_i)(x-x_i)+\frac{1}{2}f''(x_i)(x-x_i)^2$$

is determined by the remainder, we can bound the error as

$$|f(x)-f_h(x)|\le \frac{1}{6}\max_{\xi\in[x_i,x_i+2h]}|f'''(\xi)||x-x_i|^3$$
$$\le \frac{1}{6}\max_{\xi\in[a,b]}|f'''(\xi)|(2h)^3$$
$$=\frac{8}{6}\max_{\xi\in[a,b]}|f'''(\xi)|h^3. \qquad (1.20)$$

Integration of $f_h(x)$ over $[x_i,x_i+2h]$ by Simpson's formula is exact if we calculate without loss of decimal places. So the inaccuracy in the value of the integral of $f(x)$ over $[x_i,x_i+2h]$ is determined only by the remainder term. Making use of (1.20), we see that the error in the integral itself over the interval $[x_i,x_i+2h]$ cannot exceed

$$\frac{8}{6}\max_{\xi\in[a,b]}|f'''(\xi)|h^3\cdot 2h=\frac{8}{3}\max_{\xi\in[a,b]}|f'''(\xi)|h^4.$$

Because this inaccuracy happens on all n partition segments and $2nh = b-a$, the total inaccuracy of the integral over $[a, b]$ is no more than

$$n \cdot \frac{8}{3} \max_{\xi \in [a,b]} |f'''(\xi)| h^4 = \frac{4}{3}(b-a) \max_{\xi \in [a,b]} |f'''(\xi)| h^3.$$

Thus, the error in Simpson's rule for a function having a continuous third derivative is no more than ch^3 for small h, with

$$c = \frac{4}{3}(b-a) \max_{\xi \in [a,b]} |f'''(\xi)|.$$

We say that the method has order of inaccuracy h^3.

In particular, from the above, we can state that the better the local approximation of $f(x)$ by the second-order Taylor expansion (that is, the smaller the value of $\max |f'''(\xi)|$ on each of the intervals $[x_i, x_i + 2h]$) the smaller the error in Simpson's formula on each interval and thus on the whole of $[a, b]$.

We can continue finding local approximation formulas, requiring them to be precise for polynomials of third or higher order. This means that we approximate the function by a polynomial of the corresponding order. In this way, we get integration formulas of the so-called third and fourth order. In practice, the latter are normally used with nonuniform partitions. The accuracy of each approximation is given in a way similar to that of Simpson's rule, and this means the approximation is good when the integrand is sufficiently smooth.

The order of accuracy for each of these formulas is given in terms of the step size h. When we say a method is "of order" h^k, we mean that the error is bounded by ch^k for some constant c. But this constant could be so large that h would have to be taken exceedingly small. Practitioners must therefore resort to another method of verifying that a certain degree of accuracy has been achieved. The usual procedure is to perform the calculation first with step size h and then with step size $h/2$; if the difference between the two resulting values is small enough, one may stop. Such approaches are not foolproof, however. Given a desired accuracy, for example, one may think it possible to produce a finite sum that approximates the value of the nonconvergent series $\sum_{n=1}^{\infty} 1/n$.

We should also note that uniform partitioning of the integration interval does not always work well. Some functions vary slowly over certain ranges of their arguments but rapidly over others; it is evident that smaller step sizes should be used to integrate over the latter ranges. Modern integration routines often incorporate *adaptive* schemes in which the step size is changed automatically when necessary.

The inherent difficulties of numerical integration are compounded when we consider the evaluation of multidimensional integrals. For a plane rectangular region, the subdivision of each side into n parts will yield n^2 rectangular subregions. The situation becomes even worse (n^3) for a cube, and so the approximation formula for an elementary cube must have accuracy better than third order, at least. Large-scale calculation can require a great many elementary computations and introduce substantial roundoff error.

We should add that sometimes the integrand values are derived from measurements and, in this way, empirical inaccuracy is introduced into the integrand in advance.

There are other approaches to the numerical approximation of the definite integral. One is due to Carl Friedrich Gauss (1777–1855), who proposed to find the distribution of points x_i on the domain of integration for which a linear combination of the integrand values at the x_i presents the best (in a certain sense) approximation to the integral.

1.13 On the Notion of a Function

What is mathematics? Nearly every famous mathematician has tried to answer this question. But mathematics involves so many objects, relations, and tools that no simple answer seems adequate. Moreover, the content of various sciences falls partly inside and partly outside mathematics. Occasionally someone will try to fill the void with a statement that sounds almost like a cruel joke:

> Mathematics is the art of dealing with undefined objects having strictly defined properties.

Unfortunately, some nonmathematical sciences fit this definition as well. (And others fail to do so only because the properties of their objects of study cannot be strictly defined.)

We cannot even define the term "integer" in a satisfying way. We can do a lot with integers, though, and the situation is similar with points, lines, planes, spaces, sets, and so on. The properties of these objects have been strictly formulated, but the objects themselves cannot be rigorously defined: the best we can do is accept these notions a priori, just as we must accept our own innate reflexes.

Another primary notion is that of *function*. At one time, people equated it with the notion of *formula*, that is, an explicit rule that could be used to compute values, and this viewpoint led to certain assumptions that modern mathematicians would deem incorrect. For example, in some old papers, one can find proofs in which an author assumed that if a function is defined on some interval then it is automatically defined everywhere else. *Analytic* functions do have this property, and early workers had these largely in mind. Mathematical language evolves as does any other, and one must remember this when attempting to read the classic works.

Modern students learn a "definition" of function that goes as follows:

> A *function* is a correspondence between two numerical sets $D(f)$ and $R(f)$, called the domain and range of $f(x)$, respectively. To any point of $D(f)$ there corresponds a unique point of $R(f)$.

Let us subject this to a bit of scrutiny. First, what is a correspondence? "A relation," one might answer. Then, what is a relation? Someone with a good thesaurus might be able to take this dialogue pretty far, but would soon run out of words and be forced to say something like, "A relation is a function." So the circle eventually closes and we have no definition. The reader who tries to "define" anything rigorously will run into the same problem. We use a great many undefined notions in everyday life, so we should not deal too harshly with mathematicians for having to do the same thing. But we can manage to identify the crucial property of a function $f(x)$: any x we take from $D(f)$ is paired with one and only one element of $R(f)$. So we come to think of a function $f(x)$ as a *triple* consisting of $D(f)$, $R(f)$, and the law according to which y from $R(f)$ is paired with x from $D(f)$.

It is tempting to think of a function as specified by the pairing law alone. However, if we change any element of the triple discussed above, then we create a new function. Take, for example, the formula $y = \sin x$. When we specify that this is to apply for all x (i.e., that $D(f)$ is the whole real axis), then we have a function; however, if we restrict x to some interval (a, b), then we have a different function (though described by the same formula). The law of correspondence is an important part of the definition of any function, but no more important than its domain and range.

We should also note that in modern mathematics, the domain and range of a function need not be numerical sets. A set can, for example, consist of elements that are functions themselves. Functions that map functions into other functions play key roles in mathematics. They are often called *operators*, or *mappings*. We should also mention the idea of a *multivalued function*: a correspondence under which each point in a domain set can be "paired" with finitely (or even infinitely) many elements from another set. Consider a row of houses, for example. We could think of pairing the street address of each house with the house itself *along with all of its contents* at a fixed time. The information organized in this way could be useful for constructing the next population census. Such extensions of the function concept are necessary in many applications.

1.14 Differential Equations

The deflection of a certain point of an ordinary clothesline is determined not only by the item hanging at this point, but by everything hanging on the line. At first glance, this appears to involve a correspondence that could not be called a function. It turns out that the deflection of the rope can be represented as the solution of an equation involving the function itself, its derivatives, and certain conditions at the rope ends. Such an equation is called a *differential equation*, and the end conditions are called *boundary conditions*.

To show how differential equations arise in other situations, let us consider a couple of problems.

Suppose a lake contains 1,000,000 fish, and we know that this population will increase by 0.01% per day. What is the resulting law of population growth if this situation persists? Of course, we could ask the nearest bank accountant; the situation with interest rate calculations is closely analogous. But let us find the law ourselves.

When modeling realistic problems, we often simplify and idealize in order to obtain realistic results. Here, we shall suppose that the population grows continuously. This means that, during some short time period, our model may predict that an additional 0.344 fish will be added to the population. Such a thing obviously cannot occur in nature, but this is the small price we pay for using a continuous model. We also neglect any possible effects due to seasonal variations in the environment, and so forth. We even ignore differences between male and female fish. Let us assume that the process of growth at any moment depends only on the present number of fish and the time increment over which the growth is examined.

Let y denote the number of fish at time t, so it is a function $y = y(t)$. It is reasonable to suppose that the more fish we have, the greater the population increase during some time interval. Moreover, it is reasonable to assume that the population growth over a small time increment is proportional to this increment. (So, in banking terms, we would like to consider a continuous problem for compound interest.) Thus, introducing the increment of fish Δy over time $(t, t + \Delta t)$, we write

$$\Delta y(t) = ky(t)\Delta t,$$

where k is a proportionality coefficient. Here, k is unknown, and may not even be constant with time. It may depend on y as well. The validity of any assumption could only be verified by comparing the output of our model with empirical data from a real natural system. This equation can be rewritten as

$$\frac{\Delta y(t)}{\Delta t} = ky(t),$$

and, on the left, we see a now familiar structure ready for the limit passage as $\Delta t \to 0$. If this limit exists, the left member passes over to the derivative of $y(t)$. Assuming k depends on Δt continuously, we get

$$y'(t) = ky(t). \tag{1.21}$$

This is a typical ordinary differential equation; because the unknown y appears with one derivative operation applied to it, we say that the equation is of *first order*. The solution process is facilitated by taking k to be a constant, and yields a unique result when the value of y is specified at some instant t_0. Note that (1.21) describes a system with a discrete set of values, using continuous functions. Also note that we met the same equation (and the same derivation) when we modeled the population growth of a town. This is

typical in mathematics: very different systems may be described by models
that share the same abstract form.

We are familiar with Newton's second law:

$$F = ma.$$

Remembering that the acceleration a is the second derivative of the distance
function $s = s(t)$, we have

$$s''(t) \doteq F/m.$$

If we happen to know F as a function of time t (and possibly of s and s'),
then we have a differential equation of *second order* governing the unknown
function s.

We can point to other differential equations. For example, if we know the
velocity v of a point as a function of time t, then we have the relation

$$s'(t) = v(t). \tag{1.22}$$

We can consider this as a differential equation for the unknown distance s.
We also know the solution:

$$s(t) = \int v(t)\, dt.$$

If the motion starts when $t = 0$, then this information can be incorporated
by writing

$$s(t) = \int_0^t v(u)\, du + C,$$

where C is an indefinite constant. To pin down C, we need to know where
the moving point is at the initial time $t = 0$. If this initial location is $s = s_0$,
then we often write

$$s(t)\big|_{t=0} = s_0.$$

This is an *initial condition*. The differential equation and associated initial
condition serve to define the unknown function uniquely, and together constitute a *Cauchy problem*. Substituting the solution into the initial condition,
we find that $C = s_0$, and thus the solution is uniquely defined.

Another Cauchy problem consists of Newton's second law taken together
with two initial conditions, one each for the initial values of distance and
velocity. Two initial conditions are required because the solution of a second-
order differential equation will, in general, involve two indefinite constants.

In the general theory of ordinary differential equations, the Cauchy problem for an equation of the form

$$y' = f(x, y)$$

is studied. Here, the unknown y is a function of the variable x. Under some restrictions on the smoothness of f, it is shown that the Cauchy problem consisting of this equation and the initial condition

$$y(x)\big|_{x=x_0} = y_0$$

has a unique solution $y(x)$ on some interval $[x_0, x_1]$. This extends to the theory of Cauchy problems for systems of ordinary differential equations. It is important because many problems in classical mechanics can be posed as Cauchy problems.

However, it is clear that, for Newton's second law, we could specify the location of a moving point at one moment and its velocity at another. The resulting problem is also well defined, and is an example of a *boundary value problem*. When we consider a general ordinary differential equation of the second order

$$y'' = f(x, y, y'),$$

general boundary conditions may look like

$$y(x)\big|_{x=x_0} = y_0, \qquad y(x)\big|_{x=x_1} = y_1,$$

with given y_0 and y_1.

Unfortunately, a general boundary value problem for an ordinary differential equation can be unsolvable. It may also have a nonunique solution. These problems are more complex than the Cauchy problem, but of no less importance.

The situations described by differential equations can be much more complex than those described by simple functional dependences. Differential equations are employed in such diverse subject areas as the strength of materials, physics, biology, medicine, and economics. Thus we should learn how to solve them. Some approaches are analytical in nature; these often bear the name "integration" because they reduce to just that. So, the limitations on finding antiderivatives carry over to the solution of differential equations. We must therefore elaborate on the possibility of approximate solutions.

A differential equation can describe a discrete process, as we saw when considering the fish problem above. In this case, the differential equation approximated the real situation. On the other hand, a differential equation itself can be approximated because it contains a derivative that is essentially a limit of increments. The value of the limit y' can be approximated by the "difference quotient" $\Delta y / \Delta x$, and the differential equation by

$$\frac{\Delta y}{\Delta x} = f(x, y).$$

The smaller the value of Δx, the better the approximation. This approximate equation can give us a discrete sequence of y values, step by step, if we know the initial value of y.

To understand how this could be done, let us consider the simple equation

$$y' = y, \tag{1.23}$$

with $y(0) = 1$. The corresponding approximation is

$$\frac{\Delta y}{\Delta x} = y.$$

Let us take $\Delta x = 0.1$. Then,

$$\frac{y(0.1) - y(0)}{0.1} = y(0).$$

From this, we get

$$y(0.1) = 1.1.$$

The next step is to set

$$\frac{y(0.2) - y(0.1)}{0.1} = y(0.1)$$

and obtain

$$y(0.2) = 1.21.$$

In this way, we can proceed to calculate $y(0.3)$, $y(0.4)$, and so on. We can then plot points on the coordinate plane and interpolate between them using straight segments. Thus we obtain an approximate solution to the differential equation.

For this Cauchy problem, the solution is $y = e^x$. If we plot this function on top of our previous plot, we will see the difference between the two (Figure 1.8). It is not small after a number of steps have been carried out.

Perhaps our step size $\Delta x = 0.1$ was too big. If we use $\Delta x = 0.01$, we will get a better approximation for a while, but it too will eventually deviate significantly from the actual solution. The same can be said for $\Delta x = 0.001$, and it turns out that no choice of Δx is small enough to prevent this. In fact, the smaller the step size, the more steps we need to take in order to obtain the value of y at some x_0; as we do more and more elementary operations we start to see a serious accumulation of numerical roundoff error. So it is not recommended to take the step size too small or too large. There is another helpful approach, however, and that is to find a better approximation for the derivative than the straightforward difference quotient used above. Several have been devised. For example, the derivative at x can be approximated as

$$\frac{y(x + \Delta) - y(x)}{\Delta}$$

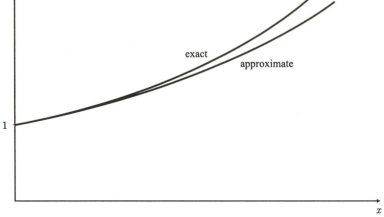

Figure 1.8 Comparison of the exact and approximate solutions to the Cauchy problem having $y' = y$ and $y(0) = 1$.

or as

$$\frac{y(x) - y(x - \Delta)}{\Delta}.$$

We can also average these to produce

$$\frac{y(x + \Delta) - y(x - \Delta)}{2\Delta}.$$

We should select the best available approximation. However, the best-
approximation problem is not uniquely posed; in various situations, we may
have different ideas about which quantity we need to approximate best of all.
Even within the same problem we may seek, say, displacements or stresses,
and these quantities will necessitate different "best approximations." Besides, it is necessary to elaborate on how we decide that one approximation
is better than another. One way is that which we used for integrals: we
can try to find approximation formulas for which differential equations having polynomial solutions up to some degree should be solved exactly. Many
useful formulas have been thus obtained. The reader should be aware that
other ideas have also been devised to approximate solutions of differential
equations in the best possible way.

Powerful and accurate methods have been developed for finding approximate solutions to Cauchy problems. It is worth mentioning that all codes
developed for this purpose can be used to calculate definite integrals. This

is because computation of

$$\int_a^b f(x)\,dx$$

amounts to finding $y(b)$ for the Cauchy problem

$$y'(x) = f(x), \qquad y(a) = 0,$$

because

$$y(t) = \int_a^t f(x)\,dx.$$

1.15 Optimization

We have considered two problems involving approximation: that of the definite integral, and that of the ordinary differential equation. In each case, an issue arose regarding optimality. We put some thought into the matter each time, but did not bother to set out precise standards that an approximation should meet in order to be called *optimal*. We could, for example, require a minimum deviation between the approximate and actual solutions of a differential equation. We would then be forced to present actual numerical evidence to back up any claim we might make about having found the best possible approximation.

The profit of a large corporation could be presented as a function of certain parameters, including the number of employees, the cost of materials and equipment, the company investments, and so forth. It is natural to seek to maximize this through optimal selection and adjustment of the various parameters involved. This is another type of optimal problem.

Many mechanics problems can be solved on the principle that an elastic body in equilibrium under external forces will configure itself in such a way that the total energy of the system is a minimum. The energy is given by an integral involving various parameters of the body and the internal state of the body at each point; for this reason, the energy is more complex in nature than an ordinary function. We see here an example of a function (i.e., the point-by-point internal state of the body) coming into unique correspondence with a pure number (i.e., the total energy of the body). Such a correspondence is called a *functional*. In short, we find the stable equilibrium states of an elastic body by seeking to minimize the total energy functional.

The problem of the minimum (or maximum) of a functional is common in modern mathematics. In a properly posed problem, the points of minimum of a functional describe the most desirable (or undesirable) situations. Many natural processes are governed by optimal principles, and whenever we formulate a problem as a problem of minimizing a functional, we say that we have chosen a *criterion of optimality*. In physical problems, it is often appropriate to work with some sort of energy functional. In areas such as economics, other factors may dictate the choice of an optimality criterion.

One of the hardest things about optimization is the initial problem formulation. It is necessary to look at a situation and devise a function the values of which will provide some measure of what is "better" or "worse" in the situation. This can be challenging, because real situations are seldom clearcut. But until we have access to a function that can yield actual numerical values, we cannot proceed further with a mathematical approach. To make matters worse, we often run into conflicting desires and requirements. The manufacturer of a certain commodity may wish to maximize her financial profits, but not at the total expense of environmental quality, her own free time, and other factors. So hard choices must often be made in order to decide which quantity is to be maximized or minimized. In the end, we cannot optimize many functions simultaneously: experience has shown that problems requiring the simultaneous optimization of several different functions or functionals are usually unsolvable. Workable optimization models for real-life problems can be formulated through a two-step procedure.

1. Select a functional that characterizes the optimal conditions, and state the problem as a problem of minimum or maximum of this functional. If several functionals must be involved, they must be somehow combined to form a single overall functional. The combination may take the form of a weighted sum or sum-of-squares, with weighting coefficients chosen to express the relative importance of each particular functional in the combination.

2. Restrict the domain over which the optimization search will be conducted. Suitable restrictions normally take the form of inequalities involving the parameters of the problem and sometimes involving the functionals themselves.

The simplest problem of optimization is the problem of finding the minima and maxima of a function on an interval. The solution of this problem is based on the notion of derivative. The first derivative permits us to identify those points at which a minimum or maximum *may* occur. Indeed, by definition of the derivative, we see that the increment of a differentiable function $f(x)$ at a point x is equal to

$$f(x + \Delta x) - f(x) = f'(x)\Delta x + \omega(x, \Delta x),$$

where $\omega(x, \Delta x)$ tends to zero faster than Δx when $\Delta x \to 0$. This means that the change in $f(x)$ at x is fully determined by the value of $f'(x)$. Thus, if $f'(x) > 0$, the values of $f(x)$ increase when we increase x; if $f'(x) < 0$, they decrease. Now, suppose we find a point x_0 such that $f'(x) > 0$ to the left of x_0 and $f'(x) < 0$ to the right of x_0. Let us further assume that $f'(x)$ is continuous so that it must pass through zero on its way from positive to negative values. As we approach x_0 along the real axis, the values of $f(x)$

increase, become "stationary" at x_0, and then decrease: this clearly looks like a maximum point. We say that

$$f'(x_0) = 0 \qquad (1.24)$$

is a *necessary condition* for $f(x)$ to have a maximum at $x = x_0$. It is obvious that the necessary condition for a minimum is the same. Some caution is required here because of our continuity assumption on $f'(x)$. The simple function $y = |x|$ has a minimum at $x = 0$ but does not possess a derivative at this point. The condition (1.24) is necessary only for functions having a continuous derivative in an open interval.

In this way, we can identify candidates for the extreme points of a function. The next step is to check whether these points are really extrema. This can be a lot of work. Minimization of a functional is even harder, and the modes of reasoning involved are more complicated. They are essential in such branches of mathematics as the calculus of variations, optimal control, mathematical programming, and functional analysis.

It is beyond our scope to consider such problems. Our real aim is to pick out some interesting real-world problems and to discuss them a bit. Optimization problems can be complex and require a long time to solve. Let us describe one important problem of optimization. Here, the exact approach will be impossible in principle, but instead we will have to solve many particular problems before we can even formulate the central one. This problem poses so many challenges that it provides a good example of a realistic interaction between mathematics and real life.

1.16 Petroleum Exploration and Recovery

In any real problem, it is possible to identify places in which mathematics is used intensively. Let us briefly consider an area of importance to us all: the location and extraction of oil from the depths of the earth. The complexity of this problem is reflected in ordinary fluctuations in gasoline prices.

Naturally occurring oil reserves usually reside deep beneath the surface: several hundred meters or more. An underground pool, which may be of complex shape, is termed a *lens*. Deep wells must be drilled in order to bring the oil to the surface, and the cost per well can be enormous.

Inside an untapped lens, the oil is under great pressure because of its depth under the Earth's surface and the presence of various gases. When the lens is first tapped with a well, this pressure is sufficient to bring oil all the way to the surface. As more and more oil is released from the lens, however, the pressure begins to decrease and eventually becomes too weak to push any more oil to the surface. A century ago, people would have been forced to abandon such a well and drill a new one. But it turns out that, for a deep lens, the quantity of oil expelled under its own pressure may amount to only around two percent of the total oil present. This may sound like a job for a pump, but viscous liquids such as oil are pumped only with great difficulty (and slowness). Another idea is to try to maintain the pressure levels inside the lens by pumping in some cheaper liquid as the oil is removed. This may sound simple at first, but an oil lens can be large and irregularly shaped. So a great many wells might have to be drilled, some for removing oil and some for inserting water. At about a million dollars apiece, say, putting in these wells can become expensive. Experience has shown that anywhere from a dozen to several hundred wells might be needed. Add to this the costs of aboveground piping, compressor stations, transportation, communications, and so forth, and you have a hefty proposition.

The first problem is how to decide how many wells to drill and where to drill them. A poor decision here could cost an investor many millions of dollars. A good decision, on the other hand, might bring oil to the surface quickly and more cheaply. So we have a real optimization problem. Unfortunately, it is not easily solved or even posed. First, we seldom know the actual shape of the lens or its prevailing physical conditions. Available information is gleaned mostly from trial wells, along with some less direct methods of geophysical exploration. But the mass of oil at a depth of 1 km could be vastly underestimated or overestimated based on these methods. Furthermore, a basic strategy is to place the water wells around the outer periphery of the lens with the goal of pumping the oil toward a well near the center. We run into bad news here, too; oil and water can interact in unexpected ways. Instead of producing an outward-propagating pressure wave in the oil, a jet of water from a water well can propagate as a "tongue" through the oil. These tongues can eventually cause problems by partitioning the lens and mixing sand and other unwanted materials with the oil that will soon be brought to the surface. Once this happens, we may have to wait decades until the oil and water-tongues shift their positions in such a way that we can again begin to pump oil profitably from the lens. Experience shows that the water-pumping approach permits the extraction of an additional ten percent of the total oil in the lens, over and above what could have been taken without this technology. If we want to increase the output beyond this, we have to pump in a special polymeric liquid instead of water. Such liquids are not cheap, though not as expensive as oil. They do, however, have some unexpected properties themselves. For example, if you tip a full glass of water, the water that leaves the mouth of the glass will certainly fall to the floor. But if you repeat the experiment with a polymeric solution and right the glass quickly enough, the polymeric solution will return to the glass. In fact, a polymeric solution behaves like a very weak resin, and attempts have been made to understand this behavior by appealing to the theory of viscoelasticity.

The use of polymeric solutions has increased the output of oil wells by 30% or so. But the formation of disastrous tongues remains possible, and therefore the potential interactions between oil and a polymeric solution should be thoroughly understood. The oil itself can be modeled as a very viscous liquid, but the whole problem remains complex because of the complex geometry and inhomogeneous nature of the lens. In mathematical terms, we require a system of nonlinear partial differential equations with the three spatial coordinates and time as the independent variables. Such a model can challenge the capabilities of even the most modern computer system.

Later, we shall consider some additional problems in which the behavior of liquids and gases must be modeled.

1.17 Complex Variables

We know that the equation

$$x^2 = -1 \tag{1.25}$$

has no real solutions. But centuries ago, mathematicians learned how to find the roots of the quadratic equation $ax^2 + bx + c = 0$. Concrete examples of this arose out of both practical considerations and general curiosity, and solutions were sought with great enthusiasm. Some of these appeared to necessitate taking the square root of a negative number, and there was a marked tendency to reject them as unreal. They were even termed *imaginary*. The square root of -1, needed to express the two solutions to (1.25), was eventually given the symbol i:

$$i = \sqrt{-1}.$$

In this way, any quadratic equation can be solved to obtain two solutions of the general form $a + bi$, where a and b are real. Numbers of this form are said to be *complex*. The study of complex numbers received a strong impetus when it was discovered that they obey the laws of ordinary (real) arithmetic; the impetus became even stronger when it turned out that many real-world problems could be solved through their use. Formulas for roots of cubic and quartic equations were subsequently established. It was shown eventually that an nth-order algebraic equation has precisely n roots, if these are counted with regard for their multiplicity; for example, the equation $x^2 = 0$ is regarded as having two equal roots $x = 0$. Our brief encapsulation of history may leave an erroneous impression that all these developments occurred rather quickly. A great many small advances by generations of mathematicians were required, and the process took centuries of effort.

Self-consistency is important in mathematics. But, in the long run, only those developments that prove widely useful can enter the mainstream of the subject. Complex numbers form the basis for defining functions of a complex variable, and the result is a theory in which differentiation and integration are done in the same way as for functions of a real variable. But the theory is powerful, in part because it allows us to describe certain two-dimensional quantities using a single variable. In a sense, the number $a + bi$ behaves much like a vector having components a and b. At first glance, the main difference between a complex number and a two-dimensional vector seems to be in the notation used: with the complex number, we omit basis vectors (although one might consider i to be a sort of basis vector, at least formally). But this seemingly small distinction turns out to be significant. First of all, complex numbers, unlike vectors, were amenable to the operation of division. It also turned out that differentiable functions of a complex variable possess smoothness properties that functions of a real variable lack, and this led to their extensive application in mathematical physics. Excitement over the complex numbers continued to grow. Because the real numbers were embedded in the complex numbers, people started to wonder whether the latter were themselves embedded in some "even more complex" number system: the idea that certain three-dimensional objects might be described by a single number was certainly attractive. Unfortunately, this was not the case. Curiously, however, it turned out to be possible for the next higher dimension: four. Sir William Rowan Hamilton (1805–1865), whose name is now familiar in mathematical physics, developed a self-consistent theory describing four-dimensional quantities called *quaternions*. This theory still finds limited application. Its rules are rather complicated, though, and our interest in things four-dimensional is confined mainly to the four-dimensional space-time continuum. Of course, the value of a mathematical structure is not *entirely* determined by its "usefulness." Some results are in common use because they are dictated by the intrinsic properties of mathematical theories. Practically useless results like Fermat's last theorem are worthy of much attention — even outside the mathematical community — because of their curious histories. Mathematics is an important part of human culture, and as such its value extends far beyond materialistic considerations.

To the early workers in complex number theory, however, issues of consistency were eclipsed by the attractiveness of certain relations that involved functions in common use. One was *Euler's formula*:

$$e^{ix} = \cos x + i \sin x.$$

This follows from direct comparison of the power series expansions

$$e^x = 1 + \frac{x}{1!} + \frac{x^2}{2!} + \frac{x^3}{3!} + \frac{x^4}{4!} + \cdots,$$

$$\cos x = 1 - \frac{x^2}{2!} + \frac{x^4}{4!} + \cdots,$$

$$\sin x = \frac{x}{1!} - \frac{x^3}{3!} + \cdots.$$

It also gives rise to *de Moivre's formula*

$$(\cos x + i \sin x)^n = \cos nx + i \sin nx,$$

by which we can extract roots. This is important because any complex number can be represented in trigonometric form:

$$a + bi = r(\cos\phi + i\sin\phi),$$

where the real numbers r and ϕ are known as the amplitude and argument, respectively.

The formal use of complex arguments in the above formulas leads to their extension to the complex arena. For a power series, the domain of convergence is symmetrical in x, and when we replace x by a complex argument z, we get a circle of convergence of some radius. So the functions above, as well as many others, can be extended to take complex arguments. Not long ago, before the function concept was clarified, it was thought that any connected portion of a function could fully define its behavior on the rest of its domain. This viewpoint arose because mathematicians had been working with *analytic functions*: that is, functions that, when regarded as depending on a complex argument, have a continuous derivative. It turns out that functions of a complex variable that are analytic in some open domain are infinitely differentiable and can be expanded in a Taylor series in some neighborhood of any point of the domain. Such functions form a rather tiny but important subset of all functions of interest.

It also turns out that, for functions of a complex variable, it is possible to introduce definite integrals along paths in the complex plane. If the integration path happens to lie in a domain over which the function is analytic, then the integral has useful properties. For example, its value depends only on the initial and terminal points of the path: this property is reminiscent of the work of a potential force. This means we can deform the contour without changing the value of the integral, as long as we stay inside the region of analyticity. Such contour integration techniques now abound in all areas of mathematical physics, from electromagnetics to plane linear elasticity and plane hydrodynamics.

1.18 Moving On

The main goal of this book is to consider how models of familiar objects were developed. For the most part, we shall maintain a continuum viewpoint and neglect the "granularity" of real atomic structure. The simple but fairly precise models that result are used in the design of buildings, bridges, automobiles, ships, and so on. Their development and use requires some mathematical sophistication along with special tools from the branch of mathematics called calculus. It was our goal in this first chapter to discuss such tools, with a few excursions into history and even into other branches of mathematics. Mathematical tools are universal: they are used successfully in all areas of physics, chemistry, biology, and the other sciences. So this first chapter could be read independently of the rest of the book, which is devoted to various problems of mechanical modeling.

Historically speaking, it is only in recent years that people have begun to pursue mathematical abstraction for its inherent interest. Much of classical calculus was elaborated for concrete purposes, and the main challenge faced by its developers was to understand and describe the laws of Nature. Of course, certain ideas, like Cauchy's notion of limit, were not applied in a straightforward manner. But Cauchy, along with many others, felt the defects of the theory and wished to achieve a certain level of perfection in their tools. In this way, mathematics moved toward abstraction and became seemingly less and less concerned with applications. But we ourselves are a part of Nature, and even our loosest mental processes can hardly help reflecting — in at least some way — the natural processes behind them.

Thus, in mathematics, many of the applicable tools were developed when people tried to solve everyday problems or to understand the universe as a whole. Methods arose and grew out of the efforts of many scientists, each having his or her own goals and ideas. These early investigators saw the "big picture" — their viewpoint was therefore in stark contrast to that presented by modern calculus textbooks, where real-world problems are presented as a side issue. Indeed, the branches of science are all interconnected. The developers of calculus understood this much better than many present-day researchers, who tend to be experts in one or two very narrow areas but completely ignorant of others. It is useful to have a general view of science. So we invite even those who have no intention of working in mechanics to continue reading this book. We will illustrate how the pioneers of mathematics and mechanics used and developed the calculus for solving many important problems of everyday life.

Chapter Two

The Mechanics of Continua

2.1 Why Do Ships Float?

We have noted that to solve the oil well problem in an optimal way, we must understand the interaction of two viscous liquids under pressure. The science of *hydrodynamics*, to which this problem belongs, has been under development for centuries. We have heard the story of how Archimedes ran naked through the streets of his town shouting "Eureka!" after he found the explanation for why water supports a body immersed in it. This is probably just a legend, similar to the apple falling on Newton's head. People love to hear stories about sudden, dramatic discoveries. But any "bolt of lightning out of the blue" is usually preceded by years of study, careful thought, uncertainty, and deep struggle over the ultimate risk one will have to take with one's reputation.

Hydrodynamics dealt initially with an ideal liquid that was incompressible and inviscid. It took time for people to understand that they could neglect many real properties of water in order to solve problems, and then more time to understand what to do with real liquids. Any scientific model of a real-life situation has to be somewhat idealized. In certain astronomical calculations, we can consider a star as a point mass even though it is enormous in comparison with the Earth. When people first considered how ships float, they used an ideal river consisting of an unbounded half-space full of water. The influences of river bank and bottom are important but represent complications that require the assistance of modern computers. So an essential step in attacking any scientific problem is the formulation of an idealized model. As with practically any other activity, we are advised to neglect small details at first and try to focus on central features. The physical picture that formed the basis of this model, in which inertia was also neglected, led to models of various other processes such as heat flow and the behavior of the electromagnetic field.

At first glance, it seems easy to define "liquid" and to state various properties relevant to the term. But even the attempt to clearly distinguish liquids from solids brings us unexpected troubles. In elementary physics, we are told that a liquid is a material that can change its shape under an infinitesimal applied force and that takes the shape of its container — a very nice "definition." Let us take a material like glass, then: is it a liquid or a solid? In Europe, there are some old church windows which, over the centuries, have become thinner near the top and thicker near the bottom. This is the direct result of a process of fluid flow. The process is slow, but it is essentially the same as what we would see occur in several hundredths of a second if we were to release a vertical "plate" of water and let gravity take over. On the other hand, when an airplane crashes into the ocean surface, the shock is extremely hard — like a collision with a stone wall. Hence, any elementary attempt at explaining the difference between solids and liquids must rely on an implicit assumption that time is measured in everyday increments like seconds, hours, or days, and not in terms of extremely long or short periods. But even then surprises can occur. A substance as familiar as water can display state changes and other behaviors in certain situations that are still poorly understood.

Nevertheless, water can be fairly well described as an incompressible ideal liquid. This model was developed over centuries, and its properties were axiomatized on the basis of a long experimental background. These properties are as follows: any volume of the liquid does not change under external pressure, and there is no friction between the liquid and either itself or other bodies. The latter means that the motion of the particles of an ideal liquid is defined only by inertia and the continuity of the liquid medium.

Inside the liquid, there can be some pressure due to gravity, surface tension, and other external conditions. The pressure p is determined as the ratio of the uniformly distributed force F acting normally on a planar domain to the domain area S:

$$p = F/S.$$

Pressure can be measured with a manometer. The invention of this device had to wait for Blaise Pascal (1623–1662), because it required an understanding of many principles of continuum mechanics. Pascal found experimentally that the pressure at any point in a liquid is independent of direction (i.e., of the way in which we orient the area S above). Today, we can verify this by submersing a manometer at a fixed point and orienting it variously to see that the reading never changes. But this property of fluids is not self-evident, and must be regarded as an axiom.[1] Pascal also established that any additional pressure acting on a closed volume of liquid is transmitted immediately to all parts of the liquid. Of course, *Pascal's law* is only a first approximation; a pressure increase must travel as a wave, and the speed of sound in water is fast but finite. A convincing demonstration of these laws and the incompressibility of water can be made by firing a bullet into a wine barrel (the wooden kind constructed with hoops) filled with water. The sudden entry of the bullet compresses the water near the entry point (the water strongly resists compression, but the bullet resists even more). The resulting pressure increment is quickly distributed to all parts of the barrel, including its surface. Because this surface is big, a considerable force acts on it; typically, the hoops cannot withstand this and the barrel explodes rather

---

[1]In physics, the word "axiom" is avoided. Instead, an "assumption" is made, but from a logical viewpoint this amounts to the same thing.

dramatically.

We would like to discuss some basic hydrodynamics with respect to its historical development. We begin with the law of Archimedes, and try to explain why a heavy ship can float on the surface of water without capsizing.

Let us note that the pressure under water can be calculated easily: the pressure on a plane surface of unit area placed at a certain depth is equal to the force the water exerts on this surface. This force is composed of the air pressure, which is approximately 1 atmosphere, and the weight of the overlying water column (this weight is numerically equal to the pressure it produces, since we are working with a unit surface area). In quiet water, the pressure is the same at all points on the same horizontal level; this holds whether or not a ship floats on the surface above. These basic laws can be elaborated into a system of partial differential equations that govern the equilibrium state of water and any bodies floating thereupon, but even without doing this, we can explain many important effects, as did Archimedes.

Let us try to explain the upward buoyant force on a floating body. Take a simple cube having face area S and let it rest upright in the water. The pressure acting on the lateral surface of the cube is self-balanced, with the same force acting on all four vertical faces; hence, the water does not shift the cube horizontally. Acting on the upper face is a pressure equal to the sum of the air pressure P and that due to the weight W of the water column above that face: the sum $W/S + P$ results in a force acting vertically downward on the cube, because pressure in a liquid is always directed normally to the surface of an immersed body. (This seems self-evident, but in a viscous moving liquid it is not the case.) The pressure on the bottom face has an additional component associated with the weight of the additional water column above the level at which this side is located. If the additional weight is W_0, then the total pressure on the bottom face is $W/S + W_0/S + P$. This results in a force acting vertically upward on the cube. The net difference in pressures between the bottom and top faces is therefore W_0/S, and because this acts over an area S, the net upward force on the cube is W_0. This is, of course, just the weight of the water displaced by the cube, and we have essentially the result obtained by Archimedes. A full proof for bodies other than cubes would involve three-dimensional integrals, vectors, and so forth, but the result would be the same: the buoyant force on a submersed (or partially submersed) object is equal to the weight of the liquid displaced.

In physics, it is shown that a distributed force acting on a body can be characterized by a *resultant* force vector, which is one of the principal characteristics of the distribution. It must be applied to a certain point in order to replace the action of the whole thing. Another important characteristic of the distributed load is its resultant moment. The two characteristics completely determine the motion if the body is rigid. For a homogeneous load such as that produced by gravity, there is a point to which we can apply the resultant force, and the moment is zero in this case. The resultant of a set of buoyant forces should be applied at the center of mass of the water displaced by the body (and no moment need be attached). This point is known as the

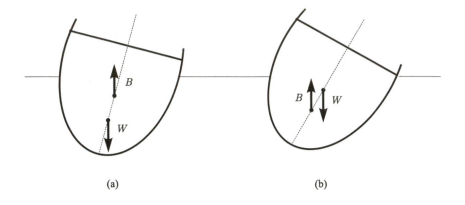

Figure 2.1 Stability regimes of a ship: (a) stable, (b) unstable. Here, W and B are the weight and buoyant forces, respectively.

center of buoyancy.

Hence, a body floats if its average density is less than that of water: when the buoyant force exceeds the weight force, the body, to achieve equilibrium, rises to the surface and reaches a position where the two forces balance. The orientation in which it floats is another matter. Powerful warships have been known to overturn upon entering deep water. Here, the problem lies in the interplay between the ship's centers of mass and buoyancy. A body's *center of mass* is a point at which an ideal support can hold the body in neutral equilibrium. It does not matter how the body is oriented about this point of support. (No ship could withstand being picked up and supported at a single point, but if we regard the ship as an ideal rigid body, then this operation makes sense on a conceptual level.) The center of mass of a homogeneous ball or cube lies at the center of symmetry. If there is an inclusion of higher density, then the center of mass shifts toward the inclusion. So, if we place heavy guns on the upper deck, the center of mass of the ship moves upward — and the shift can be considerable. Two situations can result, as shown in Figure 2.1. In case (a), the center of mass lies below the center of buoyancy. Now, if the ship is inclined a bit, we encounter a *restoring torque* that tends to counteract the rotation. Therefore, the ship has a regime of stability when nearly upright. In case (b), the same forces produce a torque that tends to incline the ship further. The greater the inclination, the greater this destabilizing torque becomes. A critical angle occurs when the two principal centers arrive at the same horizontal level. The ship then rotates to a horizontal position and water enters freely. Ignorance of these principles has been responsible for the catastrophic sinking of not a few ships, sometimes in still water (due to poor design), and sometimes in rough water (due to large waves that can set up case (b), above). An obvious way to increase stability is to position the center of mass as low as possible. At one time, this was difficult, owing to the weight of heavy masts and sails (in addition to the guns mentioned above). The only solution was to include a heavy weight, such as the main engine, in the lower portion of the ship. Some caution is obviously required when loading a ship with cargo: it can be a huge mistake to load the decks before loading the lower cargo holds! Here, as in many other areas of both engineering design and everyday life, it pays to have a solid understanding of system dynamics so that things do not get out of hand.

2.2 The Main Notions of Classical Mechanics

We have explained an important law of hydrodynamics using classical mechanics, which did not exist in Archimedes' time. For example, notions on how to add or decompose forces had to wait for Galileo Galilei (1564–1642). This makes the achievement of Archimedes all the more impressive. But now we are accustomed to the use of classical mechanics in explaining fluid phenomena.

Classical mechanics falls traditionally under the heading of physics, but it could be considered a part of mathematics because it makes intensive use of the latter's tools and approaches. In Russian universities, for example, many mathematics departments include divisions of mechanics, and students take the same general mathematics courses as do students of pure mathematics.

In classical mechanics, the main object of study is an ideal body that takes on some of the features of a real body, namely, a certain geometrical framework, along with mass distributed inside that framework. Also introduced is a rather strange object known as a *material point*: a "body" having finite mass but no spatial extent. There are situations in which only the total mass of a body determines how the body moves through space, and we can often model such a body as a material point. An example would be an approximate calculation of the gravitational effects of a distant star.

In this section, we shall take all bodies as ideal, even when discussing natural objects. The motion of a body is governed by forces of interaction with other bodies, or by "fields" of force such as the Earth's gravity. Motion takes place in space, along certain trajectories, when we treat bodies as points. The notions of space and trajectory are inherited from classical Euclidean geometry, in which, as we have discussed, the notions of space, point, straight line, and plane are taken as primitive. We repeat that we can neither define the term "straight line" nor indicate any true examples in nature (again, even a light ray coming from a distant star will be curvilinear).

To the many properties of the geometrical point, classical mechanics adds another property. This is characterized as the *inertia* of a material point: if no forces act on a material point, then it moves along a straight line. This statement forms a part of Newton's first law; the rest of the law concerns the character of the motion along this straight line. However, the law of inertia itself should be attributed to Galileo.

The branch of mechanics that describes only the geometry of motion is called *kinematics*. Classical geometry is concerned with shapes and lines, but not with how changes in mutual position occur. This is the subject of kinematics. The description of any motion is done relative to a coordinate frame attached to well-defined objects (the distant stars are useful for this latter purpose because they are so far away that they appear stationary). The idea of using a coordinate frame dates back to René Descartes (1596–1650), who founded analytic geometry by applying algebraic tools to the study of geometrical objects. It has since become one of the main foundations of mechanics.

Although it is tempting to attach a coordinate frame to the planet on which we live, we must remember that the Earth is not stationary, but rotates. Rotational motion involves acceleration, and the description of processes in an accelerated reference frame is not simple. There is, unfortunately, no way to propose an absolute inertial coordinate frame in space. We can continue to suppose that one exists, however, and this is the main statement of Newton's first law (which reformulates Galileo's statement). (The existence of one inertial coordinate frame implies the existence of infinitely many others: i.e., all those that move translationally with respect to the one assumed to exist. Of course, if we change from one frame to another, then our position and velocity values may change, but accelerations will be unaffected.) A full statement of the law is as follows: there is a spatial coordinate frame with respect to which the motion of any material point not subjected to the action of external forces will be at constant velocity along a straight line.

In fact, the existence of this frame is an axiom of mechanics. Also regarded as axioms are two other laws that Newton formulated in his *Philosophiae naturalis principia mathematica* (the *Mathematical Principles of Natural Philosophy*, written in 1687 and usually referred to as the *Principia*). We note that these laws are not due solely to Newton: they were elaborated by generations of scientists who also found many other laws governing particular mechanical motions. But Newton clearly selected and formulated his three main laws as the basis of mechanics.

Newton's second law is one of the first "strange" laws of science. It relates three quantities for an ideal object, a material point:

$$F = ma. \tag{2.1}$$

None of these quantities — the force F, the mass m of the point, or the acceleration a — lends itself to rigorous definition. Only the acceleration can be more or less well defined if we accept an unclear notion of absolute time. About force, we can merely say that it is something that makes the material point move; about mass, we can only say that it is a coefficient relating the fuzzy idea of "force" with acceleration. In fact, embedded in these notions are many other assumptions, of various natures, that we take for granted. So far, no one has seriously attempted to formulate all these assumptions in a strict logical chain. If someone were to succeed, it is probable that he or she would become obscurely famous — interest would be shown only by the philosophers, who would loudly criticize the author.

We would like to discuss the notion of force — which, although familiar, is not easy to define. Is a force something that makes a body move? Yes, but we can sit still in a chair and feel a force without moving. Perhaps it is safer to say that forces characterize the interactions of bodies. But bodies also interact chemically and in other ways not normally regarded as involving forces. In other words, it is impossible to define the notion of force rigorously. We could take the same approach as is taken with the undefined notions of geometry, such as point, line, and plane: in place of a definition, we could compose a list of properties possessed by any force. Such a list exists, but it must remain incomplete because the real forces of nature exhibit such a huge array of properties.

Therefore, a full axiomatic description of the properties of forces does not exist. There have been serious attempts, of course: Gustav Robert Kirchhoff (1824–1887), who became known for his laws describing electrical networks, wrote a book on classical mechanics, in which he attempted to replace the action of forces by the action of constraints, or restrictions, placed on motion by different obstacles. The approach received some exposure but was ultimately abandoned because of its general inconvenience and lack of clarity. Mechanicists preferred to think in terms of force despite its logical imperfections.

Newton's third law states that any force action on a body, applied at contact, has an equal but oppositely directed force of reaction, applied to the other contacting body. So forces arise in pairs. When we have a field in which a force acts on a body, we always suppose the existence of another body, the source of the field, on which there is an equal and opposite reaction.

In writing (2.1), we ignored the essential vectorial nature of force and acceleration. The vector nature of force cannot be determined logically — only through experimentation. We shall discuss it further in the next section.

It is unfortunate that Newton's second law cannot be verified on the basis of logic, and we cannot devise a test to confirm its absolute nature.[2] We cannot directly measure acceleration; hence, we cannot precisely determine mass, even though mass is defined successfully as the measure of inertia of a body. Nor, as discussed above, can we define force. The only "proof" that the law is valid is the success people have had in making predictions with it under a wide variety of conditions. It has, in fact, failed under certain circumstances, and this encouraged the development of relativity theory. But classical mechanics continues to develop as a perfectly good, self-consistent theory that describes most of the world as it exists according

---

[2]Even Newton's first law cannot be verified on the basis of logic. Early philosophers tried but failed. Their observations led them to believe that any motion in nature stops sooner or later, and thus that any body in constant motion must be acted upon by a force. Galileo was the first to seek the true relations using experimental evidence. He clearly formulated Newton's first law, and also studied the dependence between forces and accelerations due to gravity.

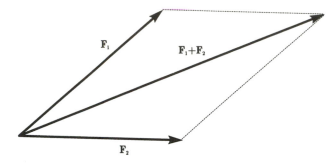

Figure 2.2 The composition (addition) of nonparallel forces.

to our five senses. It is also used in the design of myriad devices, including those employed by modern physicists to investigate new phenomena.

2.3 Forces, Vectors, and Objectivity

When we place a heavy load on a table, the weight of the items will be transmitted to the ground through the four table legs. Historically speaking, it took people a long time to understand the laws under which one force could be decomposed into several constituent forces. Galileo himself wrote a paper showing that the load of a heavy rectangular stone plate splits into halves when the plate rests evenly on two supports. It was equally hard work to understand how two given forces combine into a single force of equivalent action, especially when these forces act at an angle relative to each other (Figure 2.2). Once discovered, the *parallelogram law* of force composition formed the basis for the notion of a vector.

The vector concept is now familiar, and students first encounter it in component form. The notation $(1, 2, 3)$, for instance, is shorthand for a weighted composition **a** of three mutually orthogonal unit vectors $\mathbf{i}, \mathbf{j}, \mathbf{k}$:

$$\mathbf{a} = 1\mathbf{i} + 2\mathbf{j} + 3\mathbf{k}.$$

There are rules according to which one can multiply the vector by a number, add vectors, and compose dot and cross products. All these relate to the properties of forces as they act on material points. However, forces have properties not reflected in the purely mathematical notion of a vector. Suppose, for example, that a force acts on a rigid body. We can *slide* the vector along its own *line of action* and apply it at a new point of the body without changing its effects; this is the *principle of transmissibility*. (An alternative statement is that the moment of a force is independent of the location of the force along its line of action. We shall discuss moment in a later section.) Note that if we want to *shift* the line of action in a parallel fashion while preserving the resulting motion of the body, we will have to include an extra pair of forces (or *couple*) to compensate for any additional rotational tendency we might have introduced.

Whenever we introduce a coordinate frame, we can associate with any point (x, y, z) in space a vector **r** that starts at the coordinate origin and ends at the point. The components of this *position vector* are the coordinates of the point, so the association is one-to-one. Using the position vectors **r** = (x, y, z), we can give a clear description of motion through space. Newton's second law is easily extended from the case of motion along a straight line: we now write

$$\mathbf{F} = m\mathbf{a},$$

or, because the acceleration is the second derivative of the position with respect to time t,

$$\mathbf{F} = m\frac{d^2\mathbf{r}}{dt^2}.$$

Aside from notational convenience, there is deeper motivation for expressing the equations of mechanics in noncomponent form. A force is an *objective* (or, as they say in physics, *covariant*) entity. Objectivity concerns many objects and quantities in physics: the idea is that a spatial object and its properties remain the same under certain changes of the coordinate frame. Restrictions are stipulated on the character of the motion of new frames, not on the frame transformations at each instant. The notion of objectivity is not popular among mathematicians, who prefer to talk about the closely related idea of *invariance* instead. In a new coordinate frame, our vector will be described by a new set of components and a new dependence on the space variables, but both its magnitude and direction will remain the same as before. In this way, we are forced to obtain the new set of components, with respect to a new frame, by certain simple "transformation rules." An equation in noncomponent form is not only concise, it is valid in all coordinate frames at once. The work of recalculating components in various frames can be left to a computer program. For example, if a new Cartesian frame is obtained from an old one by rotation of the axes, then the component recalculations involve mere matrix multiplication.

Many scalar quantities in physics are objective. The mass of a body is an objective characteristic of the body; we can move our coordinate origin from the center of the Earth to the center of the Moon and the mass of a given body will not change. The same can be said of the temperature of a body, the inner energy of a body due to elastic deformation, and so forth.

Besides the scalar and vector objective quantities that show up in physics, there exist objective quantities of another type: *tensors*. The components of a vector represent the vector in some basis but tell us nothing if we do not know the basis of the coordinate frame, and a similar thing can be said of the elements of a transformation matrix. These latter elements form the components of a more complex object: a tensor. The tensor idea is connected with the vector idea, but a tensor is not a vector, because its components

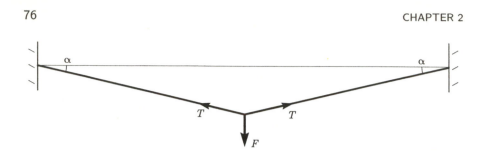

Figure 2.3 Portion of a rope under tension.

obey different transformation rules. There are many important objects of this type in mathematics and physics: inertia tensors, stress tensors, strain tensors, and so on. The importance of these objective quantities cannot be overestimated. We will return to this later, as well as to the problem of how to measure masses, forces, and weights.

2.4 More on Forces; Statics

The part of mechanics studying equilibrium of bodies under load is called *statics*. Here, one concentrates on absolutely rigid bodies that do not deform under the action of forces. However, the tools of statics can be used to describe some interesting problems involving nonrigid bodies. Take, for example, an ordinary clothesline. If we string the line too tightly before loading it with items, then it will snap when loaded. To simply blame the "initial stress" in the rope is not incorrect, but does not enable us to forecast whether the same thing would happen with another rope; the actual explanation lies in the law of force decomposition and the flexibility of the rope. Let us pursue this problem further. To simplify the discussion, we suppose the rope is loaded by a single force F concentrated at the midpoint.

Of course, we should begin by producing an ideal model of the rope. We consider it as an ideal filament that bends under the application of any force, no matter how small. We say that the rope cannot support transverse forces, hence it can only transmit longitudinal forces (i.e., forces directed along the tangent to the filamentary model at each point). How can the rope hold anything up, then? We will see that an ability to support longitudinal forces is enough. We must consider a system here, consisting of the rope and the forces that are acting on it. When this system is in equilibrium, the rope is unmovable and essentially acts to transmit force from the load to the hooks on the wall. We know that, when loaded, the rope will not remain horizontal (Figure 2.3). Its slope is symmetrical about the point of force application because that point was chosen as the midpoint of the rope. (As in elementary physics, we apply symmetry considerations without worrying too much: our ordinary experience leads us to believe that space is quite homogeneous. However, this is essentially an axiom — one of many that are seldom formulated explicitly in physical discussions.)

The nonrigidity of the rope does not prevent us from using the classical mechanics of rigid bodies. When the rope is immovable in equilibrium, we cannot distinguish whether it is flexible or rigid. Thus, for a moment, we suppose it to be rigid and apply the laws of classical mechanics. In continuum mechanics, we call this the *solidification principle*: in equilibrium, any part of a deformable body can be considered as a rigid body under the reaction forces due to the remainder of the body, and so we can subject it to the laws of classical mechanics. Applying this to the rope, we find that the tension remains constant in each portion on which external forces do not act.

Let us take only the midpoint of the rope and consider its equilibrium under three forces: the vertical gravity force, F, and the two symmetrical tension forces. By Newton's second law, the resultant force acting on the point of interest must be zero (otherwise the point would have to accelerate). The two horizontal components of tension cancel each other nicely, and it remains to relate the vertical components with F. Projection of the forces onto the vertical direction gives us the relation

$$F = 2T \sin \alpha,$$

from which we get

$$T = \frac{F}{2 \sin \alpha}.$$

This is the solution we needed. As with all physical formulas, we should examine it to see what we can learn. For small α, we will have $\sin \alpha$ small, hence T large. Therefore, an absolutely horizontal rope under a point load would experience an infinite tension and would snap. So the nature of force decomposition is what causes a clothesline to break if we try to hang it close to horizontal before we load it down. A more detailed explanation follows. The instant we apply the force F, the rope is extended a bit, and because of this, α increases to some small value at which the balance condition is met. If this occurs when the extensibility of the rope has been exceeded, our clothes will end up on the ground. In its initially relaxed state, a rope formed of individual fibers is easily extensible. So if it is not stretched tightly, then it can extend significantly, giving a larger α. The smaller the angle α, the greater the chance the rope will break.

A better model (giving α more accurately) must include a relationship between rope elongation and tension. This is a problem of the strength of materials, and in the simplest case, Hooke's law can be applied. Another thing we can learn from the rope problem is that we can consider any part of the rope — not just the center point — to be in a state of equilibrium. According to the solidification principle, in equilibrium we can consider even a nonrigid body to be absolutely rigid when applying the equations of classical mechanics. This is the central principle of continuum mechanics.

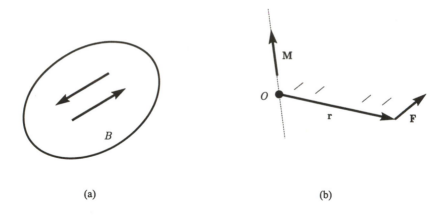

Figure 2.4 (a) A couple acting on a body B. (b) Moment of a force. The vector
   M is perpendicular to the plane of r and F.

In the rope problem, we have used the equilibrium of the center point,
replacing the rope by "reactions" from both sides. To consider the equilibrium of the rope as a whole, we must treat it as a rigid body under a load
that includes reactions at the attachment hooks. A solution based on the
solidification principle will give the same formula as before, but we should
be aware that the equilibrium states of a material point and a rigid body
differ in nature. If a force is applied to a rigid body, we are free to move
the point of application along the line of action of the force. This is not the
case for a force applied to a material point. For a rigid body, the law of
equilibrium looks similar to that for a point: we have

$$\sum_{k=1}^{n} \mathbf{F}_k = 0, \tag{2.2}$$

where the sum on the left, the resultant force, accounts for all forces $\mathbf{F}_k$
acting on the body. But the process of loading a solid body involves other
features as well.

   We have mentioned the concept of resultant moment, another characteristic of a distributed load. In classical mechanics, a *couple* is initially
determined as a pair of forces acting on a body in such a way that the forces
are of equal magnitude and opposite sense, and have parallel (but nonco-
incident) lines of action (Figure 2.4). A couple obviously cannot act on a
material point. The resultant force due to a couple is zero, but there is still
an obvious rotational effect on the body. So the idea of resultant force cannot fully characterize the equilibrium states of rigid bodies. For this, it is
necessary to bring in the notion of moment. The moment of a force $\mathbf{F}$ with
respect to a point O can be introduced as

$$\mathbf{M} = \mathbf{r} \times \mathbf{F},$$

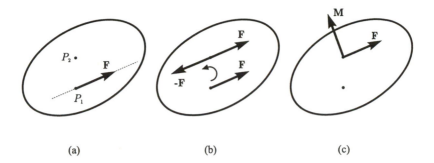

(a)   (b)   (c)

Figure 2.5 Shifting the point of application of a force **F** off its line of action. (a) Original situation. (b) Introduction of the force pair ±**F**. (c) The shifted force and couple.

where **r** is the radius vector from O to the point of application of the force. (See Figure 2.4.) In elementary physics, we learn that moment equals "force times moment-arm" and has a sign determined by the sense of rotation it would encourage if applied to a body. This is a particular case of the above formula, applicable when all forces and their associated radius vectors lie in a plane; in this case, the moment is orthogonal to the plane and must therefore be in one of two possible directions, as designated by the sign of the value. The moment of a couple, which is the sum of separate moments, each taken with respect to the same origin point, does not depend on the origin point. So we can place the moment of a couple anywhere in the body, making sure only to preserve its direction and magnitude.

A football will rotate more or less when kicked, depending on how closely the force of the kick passes through the center of the ball. This means we cannot merely shift the line of action of a force to parallel lines; we must add a compensating moment in order to predict the same motion. The procedure for doing this is illustrated in Figure 2.5. We start with a force **F** acting at point P_1 on a body, and want to move its point of application to another point P_2 not on the original line of action. We simply introduce a pair of equal and opposite forces ±**F** at P_2, and then associate −**F** with the original force **F** acting at P_1 to form a couple of moment **M**.

To describe the equilibrium of a solid body, we must impose not only (2.2), but a corresponding condition for the resultant moment:

$$\sum_{k=1}^{n} (\mathbf{r}_k \times \mathbf{F}_k) = 0. \qquad (2.3)$$

It can be shown that if (2.2) and (2.3) are fulfilled, then (2.3) is valid for any origin. The full condition for three-dimensional equilibrium of a rigid body therefore consists of six equations. For two-dimensional equilibrium problems (i.e., problems in a plane) we need three equations: two for the resultant force and one for the resultant moment.

In the dynamic case, when we are studying the motion of bodies under the action of forces, it is important to maximally simplify the system of forces. For this, we need to know how to compose forces that do not act through the same point. To change the point of application of a force in a solid body, we need to add a compensating moment, as we have said. There is an optimal line of action of the resultant force and the corresponding resultant couple to which the system of loads reduces. In the case of rigid body motion, it is done in such a way that we cannot distinguish between the effects of the original system of forces and the resultant force and couple.

We ordinarily refer to the center of mass of a body as its "center of gravity." The vertical line through the center of mass represents the line of action of the resultant gravity force, and if we place a support on this line, we place the body in equilibrium. Of course, this assumes that the gravitational field is homogeneous (which it is not). On the Earth's surface, there is also a centrifugal acceleration that affects the direction and magnitude of the resultant force.

2.5 Hooke's Law

Our discussion thus far has centered around fluids, but solids also play a central role in our lives. Some are said to be *elastic*; if we place a load on an elastic body and then remove the load, we find that the body's original shape and properties are restored. Materials such as rubber are obviously elastic, but so are a great many others. The law describing how solid bodies resist applied forces was formulated by Robert Hooke (1635–1703). Although known primarily for his law of elasticity, Hooke was an interdisciplinary genius. Among his duties as secretary of the Royal Society was presenting to fellow members two new achievements or inventions each week. He did this for years, frequently presenting the results of his own investigations. Plenty of modern scientists would be happy if they could make just one lifetime contribution on the level of any of Hooke's numerous achievements. He also discovered, for example, the cell structure of living tissues. Newton himself was obliged to Hooke for offering a correction to the first version of the law of gravitation. Robert Hooke was the Mozart of science: he introduced new ideas in many areas, but he did not write voluminous reports and his notes were often dissolved anonymously into the works of others.

Hooke discovered a basic proportionality relationship between the stress and strain in an iron wire. Stress is applied to the wire by placing it under tension, and strain is measured according to the fractional elongation that results. Suppose the wire has length L and cross sectional area S; it is found that

$$F = ES\frac{\Delta L}{L}, \tag{2.4}$$

where F is the applied force, ΔL is the elongation of the wire, and E is a constant called *Young's modulus*. The value of E depends on the wire material. We can stretch a typical iron wire until the ratio $\Delta L/L$ reaches a few thousandths, and after that it snaps. It is clear why the average person may not regard materials such as iron or steel as capable of elongation.

Rearranging Hooke's formula as

$$\Delta L = \frac{FL}{ES},$$

we see that if L is large, then ΔL can be measured without tools of great precision. A plot of F versus $\Delta L/L$ yields a straight line having slope ES.

Equation (2.4) can also be rewritten as

$$\frac{F}{S} = E\frac{\Delta L}{L}.$$

This time on the left-hand side, we see the expression used for defining pressure: the ratio of a force to its area of application. In elasticity, this is called the *stress* and is denoted by σ. Introducing the *strain* as $\varepsilon = \Delta L/L$, we have

$$\sigma = E\varepsilon. \tag{2.5}$$

It turns out that this law holds quite well in diverse situations. Materials possess a *proportional limit*; as long as we do not exceed a certain strain, a linear stress-strain relationship is preserved. It should be said that some materials exhibit nonlinear relationships and some are inelastic even for small strains. The behavior of other materials under load depends on the history of loading, and for these it makes no sense to even plot stress versus strain. But steel, wood, aluminum, and other elastic materials find such wide use that it is worthwhile to consider the question of dependence more carefully. We should mention, however, that although the linearly elastic model is good for many materials, it is always approximate: within some higher accuracy, any material exhibits nonlinear and inelastic properties.

Hooke's law can be applied to springs as well as to wires: the elongation x of a spring is related to the applied force F by the linear equation $F = kx$. Further examples would reveal that this form of Hooke's law, in which deformations are considered finite, suffices for most stretching problems. In order to treat more complicated deformations, we must reformulate the law in terms of infinitesimal quantities. This is because, in a combined deformation, each small portion of a material can exist in its own special circumstance. As we face this potential nonuniformity, we are reminded of how we handled nonuniform motion: we were forced to either describe the motion in average terms, or apply the derivative to get local information.

It is clear that the law of deformation is based on the ratio $\Delta L/L$ that we called strain. Let us see what happens when we take L smaller and smaller. First, we introduce a longitudinal coordinate x along the wire axis

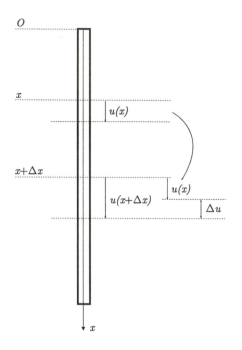

Figure 2.6 Derivation of the expression for local strain in a wire.

and measured from the end of the wire. (We continue to use the term "wire," but in textbooks on the strength of materials, in which such problems are considered, they call the object a "rod.") Let us consider a small volume between the cross sections having coordinates x and $x + \Delta x$, as shown in Figure 2.6. Note that as the total length L we exploit Δx, which corresponds to the notation we used when we introduced the derivative earlier. Some force acts on the wire that can be distributed along the length. (To see how the strain can be introduced, it is helpful to think of a heavy column, instead of a wire, hanging by one end. A given cross section of the column will be loaded with a force equal to the weight of the portion beneath that cross section.) Let us consider a function $u = u(x)$, giving the displacement of the cross section whose coordinate is x. The displacement of the cross section $x + \Delta x$ is $u(x + \Delta x)$, and the change in length of the piece Δx is $u(x + \Delta x) - u(x)$. For this piece between cross sections, the average ratio $\Delta L/L$ takes the form of a difference quotient:

$$\frac{u(x + \Delta x) - u(x)}{\Delta x}.$$

The limit passage $\Delta x \to 0$ gives us the strain at point x:

$$\varepsilon = \frac{du}{dx}.$$

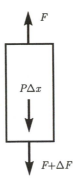

Figure 2.7 Positive directions for the forces on a wire segment. P is the density per unit length of an applied load.

An extension of this expression for ε, the so-called strain tensor for small displacements, can be used to generalize (2.5) to the case of nonuniform loading of a wire. Later, we will see how this can be extended to describe spatial deformations of a body.

Hooke's law is an example of a *constitutive relation* for a material. To solve an equilibrium problem in continuum mechanics, we must have, in addition, the relevant equilibrium equations that follow from the classical mechanics of nondeformable bodies. In classical mechanics, it is common to partition a system and consider only a certain portion; the effects of the rest are taken into account through the inclusion of unknown reaction forces. This method, together with the solidification principle, is also used for deformable bodies. In the case of a wire, we believe that the cross section after deformation remains planar (in fact, this assumption was built into the way we introduced strains). If we cut the wire at some point x, the action of adjoining parts on each other is given by equal and opposite forces distributed uniformly over the cross section. If we cut out a small piece of wire between x and $x + \Delta x$, we see that three forces act on it: the reactions F and $F + \Delta F$ from the two cutaway portions, and a force distributed over the length (an example of the latter is weight, the resultant of which can be approximated as $P\Delta x$ over a small piece Δx). In equilibrium, by the solidification principle, this piece of material under load can be treated as a rigid body and so the equilibrium equations of classical mechanics hold for it. Because the forces occur in pairs, it is necessary to appoint a positive direction for them. In Figure 2.7, we show the forces directed positively: we consider them positive if they stretch the small piece, and negative if they compress it. The resultant vertical force in equilibrium should equal zero:

$$F + \Delta F + P\Delta x - F = 0.$$

It follows that

$$-P = \frac{\Delta F}{\Delta x}.$$

Dividing both sides by the cross sectional area S, we get

$$-\frac{P}{S} = \frac{\Delta \sigma}{\Delta x}.$$

Denoting $p = P/S$ and passing to the limit as $\Delta x \to 0$, we get the equation of equilibrium in the form

$$\frac{d\sigma}{dx} = -p. \tag{2.6}$$

Substituting σ from Hooke's law (2.5), we get the equation of equilibrium in terms of displacements:

$$E\frac{d^2 u}{dx^2} = -p. \tag{2.7}$$

This is a second-order ordinary differential equation. To pose the problem properly, we should appoint boundary conditions. For example, we can suppose that the point $x = 0$ of the wire is clamped,

$$u(0) = 0, \tag{2.8}$$

and that a force F_0 acts on the other end:

$$ES\frac{du}{dx} = F_0. \tag{2.9}$$

The three equations (2.7)–(2.9) constitute a boundary value problem for u. It can be solved without integration if the law $p = p(x)$ is piecewise linear. We will not discuss particular problems, as they can be found in any textbook on the strength of materials.

At this point, we shall apply the simple law (2.5) to a slightly different problem.

2.6 Bending of a Beam

We now come to one of the main problems of the strength of materials. Our model of how a beam bends dates back to the work of Daniel Bernoulli (1700–1782) and Leonhard Euler (1707–1783), a pair of great mathematicians and mechanicists.

One cannot understand the bending of a beam by logical reasoning alone; much experimentation with real beams was necessary to elaborate the main

Figure 2.8 A beam in bending: nondeformed and bent states.

ideas involved. Early mechanics was based on the idea of linearity — that any function can be approximated by a linear function on a small domain of its parameters — and there is a corresponding theory covering the bending of thin beams (or *elastica*, as Euler called them). Although the linearized theory would do a good job at describing, say, a long, thin ruler, it would start to fail as applied to thicker and thicker structures.

Let us take a long, thin, sufficiently wide, and straight ruler whose cross section is a rectangle with sides h and b. To qualify as thin, the ruler needs an aspect ratio h/b of much less than 1, say, $1/10$ or smaller. Draw the midline on the lateral surface and a few normals to that line at several points, as shown in Figure 2.8. Bending the ruler a bit by applying moments to the ends, we see that the midline bends in such a way that its length practically remains unchanged; furthermore, the normals rotate a little and remain normal to the midline in its new position. These assumptions become the main postulates for Bernoulli's model, and also apply to the theory of plates and shells (where they give rise to *Kirchhoff–Love models*).

Hence, we assume that the beam, when bending so that all of its points move parallel to a plane (the *plane of bending*), has a midline that remains the midline in the bent state, and that the normals to this line remain unchanged in length and normal to the bent midline. These bending kinematics permit us to derive a workable model of a beam.

As for the earlier problem of stretching a wire, we will see that the system of relations splits into two parts: one describes the statics of the process, which is not concerned with the physical properties of the beam material, and the other relates statical characteristics with the deformation through Hooke's law. We begin with the statics of an elementary part of the beam (ruler). Let the beam be deformed by some transverse load. Our derivation will assume that the deflection y of the midline at all points is small, as are all rotation angles θ of the normal lines. Of course, no length can be declared small in an absolute sense. So we should talk about relative smallness instead, requiring $|y|/h$ to be small over the ruler length. But it is common in engineering and applied science to perform derivations as if such quantities were extremely small (infinitely small in fact); this allows us to neglect higher powers of these quantities (such as $(y/h)^2$). We also replace $\tan\theta$ and $\sin\theta$ by θ on the basis that θ is infinitely small. In practice, the resulting formulas are applied to situations in which $|y|/h$ gets as big as 10 for small h/L, where L is the length of the ruler. Such extension of the range of applicability of a formula, well beyond the conditions under which it was derived, must be based on experiment. It is, however, a common situation in physics.

The general problem of small deformation under any load splits into two problems: bending and stretching. The latter problem was solved in the previous section. The force reactions for the stretching problem do not participate in the bending equations (the reader can verify that they participate only in the equation for moments, but we will take the origin point for the moments on the line of action of the resultant stretching reactions, so they will have zero moments).

So, let us consider a piece of ruler between two closely spaced normals to the midline after deformation (the discussion is given in terms of the longitudinal cross section, but the forces and moments are those that act on transverse cross sections of the ruler). Because we are interested only in bending, the reaction of the rest of the ruler can be replaced by a vertical *shear force Q*, a horizontal force that we can neglect at bending, and the moments. The reaction forces and moments are distributed over the ruler cross section, but in the model under consideration, it is enough to consider only their resultants. In the strength of materials, we use the resultant characteristics of the stress state, deciding on the points of application of forces by using symmetry and other "common sense" reasoning. But sometimes such an approach fails, and we see yet another published paper entitled "On the errors of somebody's model of . . . ."

Newton's third law states that all the forces occur in pairs, and this means that when we cut off some part and replace it by a reaction, an equal but opposite reaction should be applied to the other part. We should appoint positive directions for the reactions. For a horizontal ruler, the left and right sides are in different positions with respect to the cut parts, and this means that the positive directions from the left and right should be opposite. We take as positive the directions shown in Figure 2.9.

When deriving equations in the technical sciences, we use a lower standard of rigor than that used in calculus. Of course, we could do everything with ε-δ proofs, but things would be lengthy and intuitively unclear. Each technical science, such as mechanics, has developed efficient shortcuts that give correct results. For example, a common situation in classical mechanics is when a point moves through a small portion of a circular arc. The usual technique is to replace a small arc of the trajectory by a small segment of the tangent line. In this way, we neglect all the values of the second order of smallness of the arc angle, but get correct results in a simpler way than we would while trying to use exact values for the length of the arc and its curvature.

A frequently used trick in the strength of materials and the linear theory of elasticity is the following. The equations of equilibrium and the rest of the reasoning are derived using the initial picture of the body, the nondeformed picture. This makes a thoughtful student uneasy. It is clear that a piece of

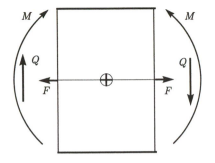

Figure 2.9 Sign conventions for a section of beam. The circled positive sign indicates that the directions shown are for positive values of F, Q, and M.

a body under consideration is in equilibrium (with use of the solidification principle) only *after* deformation, and it would be much more sensible to use the latter state in our reasoning. However, the equations of linear approximation, when we derive them from the deformed state and retain only terms of the first order of smallness, are the same as those obtained from the undeformed state. One reason for this is that the projection of a force onto some direction after rotation through a small angle α (this smallness is a necessary assumption of a linear theory) differs from the original value by a factor $(1 - \cos\alpha)$, which is equivalent to $\alpha^2/2$, and so the difference is of second order. Let us repeat the main idea of the linear approach: for the description of deformations and forces in equations and boundary conditions, we will use the undeformed state of the body as the reference state (configuration). After solving the resulting linear model for the deformations, we can "apply" these deformations to the body and get a picture reminiscent of the emanation of a "soul" from a cartoon character. It might seem more reasonable to do otherwise and, say, consider the air pressure on a body as it is shaped *after* a deformation occurs — but the resulting problem will be nonlinear and solved only with great effort (in large part, numerically). The linearity assumption also opens the door to application of the superposition principle. We can solve a problem separately for two different sets of applied loads, then combine the resulting deformation pictures to obtain that for a composite load.

Thus, we apply this famous trick (which, by the way, has been used since at least the time of Bernoulli and Euler) of using the nondeformed state.[3] The picture of the reactions and distributed loads on a part of our ruler (of small length Δx) is given in Figure 2.10. Note that we have dropped

---

[3]Pure mathematicians have always frowned upon such assumptions. They are often able to put things on a rigorous foundation, as with the "delta function" originated by physicist Paul Dirac (1902–1984). The explanations of mathematicians are unfortunately often so complex that few physicists can comprehend them. Physicists can find comfort in the existence of these occasional justifications nonetheless.

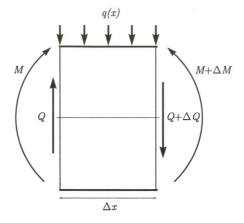

Figure 2.10 Small section of beam for analysis of equilibrium.

the longitudinal forces (we consider only bending; hence, suppose that there is no longitudinal external force and that in the notation of the previous section, we have $F = 0$).

The left side of the rectangular piece has coordinate x, so the right side has coordinate $x + \Delta x$. The reactions from the left and right sides differ by some small quantities because of the distributed load of density q over the top side of the piece. Note that the positive direction for q is down, as is that for the right reaction $Q + \Delta Q$. The density $q(x)$ is supposed to be variable; however, here we use another trick from the strength of materials, in which we replace it by a point force of amplitude $q(x)\Delta x$ applied at the midpoint of the top side (no moment is included). An explanation is that if we try to calculate the real resultant force on the small part Δx, we get an error of the second order of smallness with respect to Δx. The resultant moment of the distributed $q(x)$ is also of the second order of smallness, and so is neglected.

Using the solidification principle, we write out two equations: the sum of projections of all forces on the vertical direction is equal to zero, and the sum of all the moments with respect to the center of the left side of the rectangle is equal to zero (we recall that, in this, we will use the resultant force of the distributed load). Note, in addition, that the principle holds for the deformed state of the body, but it seems that we apply it for the initially nondeformed state. This is correct, because it introduces errors of a higher order of smallness into the equations, and so we obtain correct results within our chosen accuracy. Thus, the equilibrium equations are

$$-Q + q(x)\Delta x + Q + \Delta Q = 0$$

for the vertical forces, and

$$-M - (q(x)\Delta x)\frac{\Delta x}{2} - (Q + \Delta Q)\,\Delta x + (M + \Delta M) = 0$$

for the moments with respect to the point x. From the first equation, it follows that

$$\frac{\Delta Q}{\Delta x} = -q(x),$$

and thus the standard limit passage brings us

$$\frac{dQ}{dx} = -q(x). \tag{2.10}$$

The second equation can be rewritten as

$$\frac{\Delta M}{\Delta x} = q(x)\Delta x + Q + \Delta Q,$$

and after the limit passage as $\Delta x \to 0$, it brings us to

$$\frac{dM}{dx} = Q \tag{2.11}$$

(of course, we suppose that here Q changes continuously, and thus $\Delta Q \to 0$ as well). Equations (2.10) and (2.11) are the differential equations of equilibrium of the ruler, and do not depend on the constitutive law for the material. Differentiating (2.11) and substituting Q from (2.10), we derive

$$\frac{d^2 M}{dx^2} = -q(x). \tag{2.12}$$

Note that we did not suppose external point forces and moments on the ruler; in points of application of lumped forces and moments, Q and M have jumps.

Here we have met another point that could trouble a mathematician. The smallness of quantities under consideration is relative, and a more rigorous discussion would require the introduction of dimensionless quantities describing the system. In this case, it could be a new independent variable $\xi = x/L$; the result would be the same, however, so in the strength of materials, discussions are given in terms of dimensional quantities.

Finally, we should note that the above assumptions can be verified. This can happen (1) experimentally for some special cases; and (2) theoretically, by solving corresponding problems within the framework of a theory having a higher degree of accuracy, which is regarded as an exact one.

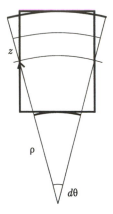

Figure 2.11 Derivation of the constitutive equation for a ruler in bending.

Constitutive Equation for a Bending Ruler

We have made some assumptions regarding the cross sections of our bent ruler: they remain undistorted and perpendicular to the deformed position of the midline, which bends but maintains its length. This picture is found experimentally by bending a ruler with two equal moments applied to the ends: in each point, the ruler has the same state (the bending moment, the only load on the ruler, is the same over all cross sections). This goes along with an intuitive idea that we can achieve the same state at each point in a bent ruler only if the ruler (i.e., its midline and every line parallel to it) follows an arc of a circle. In deriving the constitutive equation, we will use the same assumption, and we will even apply the result to cases in which the load is such that the ruler is not deformed into part of a circle; our justification for this will be the fact that each segment of the ruler can be considered locally circular, and departures from this bring only small errors.

The mechanism of bending (i.e., the constitutive relation between stress and strain) can be explained purely in terms of the moments and longitudinal strains in the ruler. We will show how this can be done (Figure 2.11). We must first derive a formula for the strains. Let us begin with the elementary formula for a circle relating the central angle (in radians) θ to the radius ρ and the subtended arc l:

$$\theta = l/\rho.$$

The same holds for an infinitely small angle $d\theta$, so $d\theta = dl/\rho$. Now we introduce another trick: for calculation purposes, we can replace an infinitely small arc of a circle by its chord, and consider the lengths of these two curves to be the same (the resulting error is of second order and can be neglected). Because by hypothesis, the length of the midline remains the same, instead of dl we can use dx everywhere in relations of the first order of smallness. In particular, $d\theta = dx/\rho$. Thus, by the above assumption, the midline remains unstretched, but all lines parallel to it are stretched or compressed, depending on their positions (Figure 2.11). Let us find this relative extension (the strain). We introduce a vertical z-axis. Each line parallel to the midline becomes a circle whose distance from the midcircle, the deformed position of the midline, is z, and whose radius of curvature is $\rho + z$; furthermore, because the small angle $d\theta$ remains the same, the length of the corresponding arc becomes

$$(\rho + z)\, d\theta.$$

The difference between the old and new states of this line is $z\, d\theta$, and thus the strain ε, which is the ratio of the extension to the initial value of the arc length dx, is[4]

$$\varepsilon = z\frac{d\theta}{dx} = z/\rho.$$

We see that the longitudinal strain varies linearly with z, the distance from the so-called neutral axis. The strain is positive for $z > 0$ when $\rho > 0$; this indicates that, if the ruler were composed of longitudinal wood fibers, those fibers lying above the neutral axis would then be lengthened. Similarly, those fibers lying below the neutral axis ($z < 0$) would be shortened. In any calculus textbook, the reader can find the formula for the curvature $d\theta/dx$ of a function $y = y(x)$, which is

$$\frac{1}{\rho} = -\frac{y''}{\sqrt{1 + y'^2}}.$$

Let y be the deflection of the midline. Because in the theory of bending of beams we suppose a smallness of both deflections and their derivatives, we can use the approximate expression $\sqrt{1 + y'^2} \approx 1$, so

$$\frac{1}{\rho} = -y''.$$

Thus, the distribution of ε over the cross section is

$$\varepsilon = -zy''. \tag{2.13}$$

So, by Hooke's law, the distribution of the stress normal to the cross section is

$$\sigma = -Ezy''.$$

---

[4]We have stopped using finite increments and have begun to use differentials, in terms of which we get derivatives directly "without" use of limit passages. This is also a common practice in physics: it reduces some of the formal effort and yields the same result as we would obtain using increments and limit passages.

The resultant moment of this distributed normal stress over the cross section is

$$M = -\iint_S E z y'' z \, dy \, dz,$$

and gives us the same integral moment we have introduced above. Here, we integrate over the cross sectional domain S and get the integral

$$\mathcal{I} = \iint_S z^2 \, dy \, dz,$$

called the *moment of inertia* of the cross section S (about the bending axis). With this notation, we get the constitutive equation for the ruler in the form

$$M = -E\mathcal{I}y'', \qquad (2.14)$$

which is the final result of our efforts in this subsection.

If we consider a cantilever beam under load, this equation is enough to find the deflection function by simple double integration, because we know the initial value for y ($y(0) = 0$, $y'(0) = 0$) and M as a function of x. In cases where there is clamping at other points, we should bring in the equilibrium equations we have derived. Substituting (2.14) into (2.12), we get the ordinary differential equation

$$E\mathcal{I}y''''(x) = q(x), \qquad (2.15)$$

which is the basis for a large part of the strength of materials.

Together with the boundary conditions, this equation constitutes a boundary value problem for y, the deflection of the ruler (beam).

For various shapes of the beam cross section, we can make the same set of assumptions: namely, (1) the existence of a midline of the beam in the "inertia" center of its cross section, and (2) that after deformation, the cross section remains planar and orthogonal to the midline. These are not so "workable" as for the rectangular cross section case; however, they still lead to (2.15), whose solutions agree with experimental data for many cross section shapes we meet in engineering practice.

Equation (2.15) is simple. So are various other equations of the strength of materials, and the reader could wonder why books on the subject are so extensive. The answer can be given from two viewpoints. The first is historical. Throughout the long history of the strength of materials, engineers elaborated simple graphical tools for the solution of this and similar boundary value problems. These tools gave solutions via simple graphs and rules for loads that are linearly distributed, or by other laws; the calculations reduced to simple multiplication and to the presentation of diagrams for moments, forces, stresses, and displacements (the reader should not forget that computers came into the picture only recently). The second explanation is that the strength of materials describes many particular cases of deformation and various situations. Many of its formulas contain factors found experimentally. We could even say that the strength of materials is more an art than a science, a field in which an engineer should clearly understand what happens when he or she uses some of the elastic system. Suppose, for example, that we have a beam attached to some structure with three bolts. The bolts are located at the vertices of a triangle, and we would like to formulate a corresponding boundary condition. Let us first assume the support is rigid, so that the beam end is fully clamped. But where do we locate the coordinate origin for this beam: at one of the vertices, perhaps? Engineers must address safety concerns along with many other issues; the tight pins will weaken with use, allowing the beam to rotate slightly, and a design engineer should be able to forecast this situation. Finally, considering the massive beam and the rest of the structure, the engineer sees that the points of support move a bit under load, and this implies a third kind of boundary condition — one that could be determined either approximately or exactly by solving an elasticity problem involving both beam and support. These questions cannot be solved by simply saying, "let the boundary conditions be such and such, and so on." The engineer should understand which conditions, restrictions, and assumptions should be applied in a given situation. Textbooks on the strength of materials have lots of supplementary material describing experimental factors introduced into equations to get the most appropriate results, and the engineer should have a good idea of the situations in which a given model is workable. In other words, in the strength of materials, there are questions whose answers depend on engineering experience and intuition, and in this sense, the subject is an art.

The formulas of the strength of materials were elaborated by generations of engineers, and have provided a basis for the design of many practical structures. People had constructed enormous ships and buildings for a long time without the benefit of any formulas. However, these things were accomplished by talented people. Science has reduced the requirements on the talents of a designer. This happens in any branch of human activity.

In the memoirs of the academician Krylov[5] there is a story that shows the difference between the scientific and intuitive approaches. While a student at a marine school in Saint Petersburg, Krylov practiced for some time at a shipbuilding factory where the chief engineer was P. A. Titov, a talented

---

[5] Alexej N. Krylov (1863–1945) was a great engineer and shipbuilder. Before the revolution, he was a tsar admiral, and after the revolution, he continued his engineering duties in the Red Army. His top official position was as Vice President of the Soviet Academy of Sciences. Being a practical engineer, Krylov was interested in science. He translated Newton's *Principia* into Russian along with other works, wrote many papers on engineering practices, and even wrote a textbook on ordinary differential equations and the practice of numerical calculations. Among his relatives were three Lyapunov brothers — the famous mathematician A. Lyapunov, a second brother who was a musical composer, and a third brother who was a philologist — and the well-known ophthalmologist Philatov. The Nobel Prize winner P. Kapitza was Krylov's son-in-law. For more information, see A. N. Krylov, *My Memories*, Sudostroenie, 1979, 478 pp.

and self-educated man of age 49. He knew only arithmetic, but could design ships by intuition. They had mutual respect for each other, and from time to time, Titov requested that Krylov calculate the minimum safe thickness of a beam. When Krylov presented him with calculations, Titov said: "Your calculations are correct, student. You see, I decided on the same size of the beam myself here." Finding himself excited by new possibilities and the evident advantages of mathematics, Titov began to study under Krylov's direction. His two-year-long study of algebra, trigonometry, analytic geometry, calculus, statics, and the strength of materials could be called a brave achievement for a not-so-young man.

2.7 Stress Tensor

We have seen that we need five things in order to describe the behavior of a deformable elastic body: the stress, the strain, the law relating stress to strain (Hooke's law), the equilibrium equations, and the boundary conditions for the body. When modeling the wire and bent ruler, we exploited simple hypotheses regarding deformation, and therefore could describe everything using simple, one-dimensional tools. In space, the picture is more complex: one-dimensional considerations and simple deformation pictures must be replaced by information obtained on a strict formal basis through the solution of equations. But the main theoretical constituents remain the same: we must describe stresses and strains locally, relate them by something analogous to Hooke's law (i.e., constitutive equations), then bring in the equations of equilibrium or motion and the boundary conditions.

Let us begin with stresses in a deformable body. We have said that in an ideal liquid, the pressure acts uniformly in all directions. For a bent ruler, we saw a clear asymmetry of the internal forces in the vertical and horizontal directions. This suggests that the stressed state of a nonliquid body cannot be described purely in terms of pressure. In a cross section of a stretched bar, we saw longitudinal stresses. A more complicated case is a nail in a wall, supporting a bookshelf: over a cross section of the nail, we will see shear (cut) stresses and a stretching load as well. We suspect that if we take a small, arbitrary cross section in a deformed body, the reaction of the other part can have any direction and magnitude. At a given, fixed point, we could examine the reactions over cross sections having various orientations, and we would find that these vary. If these reactions were absolutely arbitrary, then we would not be able to calculate anything. Fortunately, however, it turns out to be enough to determine the reaction forces (stresses) on just three cross sections having mutually perpendicular orientations (in fact, it is sufficient to know the reactions on any three different elementary cross sectional areas); those on all other cross sections are then defined uniquely. This was discovered by Cauchy, whose method of proof has been handed down with only minor modifications. We remind the reader that we deal with a continuous model of the material, hence we neglect its atomic structure.

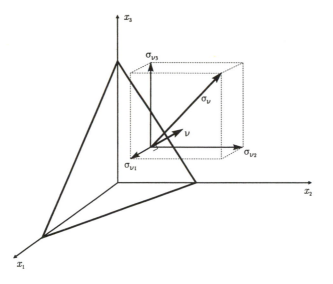

Figure 2.12 Decomposition of a stress vector $\boldsymbol{\sigma}_\nu$ into components $(\sigma_{\nu 1}, \sigma_{\nu 2}, \sigma_{\nu 3})$. The stress acts over an area element whose normal vector is $\boldsymbol{\nu}$.

Moreover, we suppose that all quantities defined inside the material depend continuously on position and, if necessary for the derivation, are continuously differentiable. There are situations in which one must consider nonsmooth functions, but we need not worry about those here.

We will consider a nonuniformly stressed state of the material, and so we deal with the deformed body in equilibrium. Over any cross section, a reaction load acts, and we will describe this (just as we did with pressure) by its density: the limit of the ratio of the force over a small portion of a plane to the area of that portion as the area tends to zero. Although the density is not constant, our use of infinitesimally small areas will allow us to treat it as such; moreover, we will replace the distributed reaction over a given area element by its resultant applied at the centroid of the element.

Unlike pressure, which is always normal to the surface over which it acts, the density of reaction stress is directed arbitrarily. Hence, it must be described in terms of components. We shall use a Cartesian coordinate system for this purpose.

Because the density vector depends on the orientation of the area element, we need a way to define the latter orientation. This is done through specifying the normal to the area element. We will therefore denote the directional dependence of the stress components by means of two subscripts: a first subscript indicating the direction of the normal to the area element, and a second indicating the axis onto which we are projecting the corresponding stress vector. Thus, if we take the stress tensor on the area element with normal parallel to the unit frame vector $\mathbf{i}_1$ (Figure 2.12), then the stress vector $\boldsymbol{\sigma}_1$ has components $(\sigma_{11}, \sigma_{12}, \sigma_{13})$. On the area element with normal parallel to $\mathbf{i}_2$, the stress vector $\boldsymbol{\sigma}_2$ is $(\sigma_{21}, \sigma_{22}, \sigma_{23})$ and, similarly, on the area element with normal parallel to $\mathbf{i}_3$, the stress vector $\boldsymbol{\sigma}_3$ is $(\sigma_{31}, \sigma_{32}, \sigma_{33})$. For an area element with arbitrary unit normal $\boldsymbol{\nu} = (n_1, n_2, n_3)$, we denote the components of the stress $\boldsymbol{\sigma}_\nu$ by $(\sigma_{\nu 1}, \sigma_{\nu 2}, \sigma_{\nu 3})$.

An internal cross section of the material is to be taken as a surface belonging to two different parts of the material, and the mutual reaction of these parts is described by the reaction stress. These mutual reactions, acting on opposite sides of a given cut, are equal and opposite, by Newton's third law (otherwise this massless surface would experience infinite acceleration). In addition, the normal directions on the opposite sides of the cut are opposite, hence we must select a positive direction for each normal. For a volume region, it is conventional to choose the outward-normal direction. So if $\boldsymbol{\nu}$ is the positive normal to the material on one side of a cut, then the positive normal to the material on the other side is $-\boldsymbol{\nu} = (-n_1, -n_2, -n_3)$. Moreover, by Newton's third law, we have

$$\boldsymbol{\sigma}_{-\nu} = (\sigma_{-\nu 1}, \sigma_{-\nu 2}, \sigma_{-\nu 3}) = (-\sigma_{\nu 1}, -\sigma_{\nu 2}, -\sigma_{\nu 3}).$$

Let us now find $\boldsymbol{\sigma}_\nu$ on the element having an arbitrary normal $\boldsymbol{\nu}$, assuming that at the same point we know the three stresses $\boldsymbol{\sigma}_1$, $\boldsymbol{\sigma}_2$, and $\boldsymbol{\sigma}_3$ on the area elements whose normals are parallel to the coordinate axes (recall that we are dealing with infinitesimal areas and volumes). We will show that

$$\boldsymbol{\sigma}_\nu = \boldsymbol{\sigma}_1 n_1 + \boldsymbol{\sigma}_2 n_2 + \boldsymbol{\sigma}_3 n_3, \qquad (2.16)$$

if $\boldsymbol{\sigma}$ depends continuously on position. For this, we will reproduce Cauchy's reasoning. It is enough to derive the formula for an origin O of a coordinate frame that is introduced into a deformed body. Cauchy considered the equilibrium of an infinitely small triangular pyramid. Three faces of the pyramid are on the coordinate planes $x_1 O x_2$, $x_1 O x_3$, and $x_2 O x_3$, and the fourth face with normal (n_1, n_2, n_3) is arbitrary, as shown in Figure 2.13. We use the above notation for the stress vectors on the faces. Note that, on the coordinate faces, the outward normal directions are opposite the corresponding axis directions. So, on the face belonging to $x_1 O x_2$, there acts a reaction stress $\boldsymbol{\sigma}_{-3} = -\boldsymbol{\sigma}_3$; similarly, a stress $-\boldsymbol{\sigma}_2$ acts on $x_1 O x_3$, and a stress $-\boldsymbol{\sigma}_1$ acts on $x_2 O x_3$.

The stresses in Figure 2.13 are force densities; into the equilibrium equations, however, we must substitute the forces themselves. So let us calculate the resultant force for each of the faces. For this, we must introduce the areas of the coordinate faces: we call them S_1, S_2, S_3, respectively. The area of the inclined face will be called S_ν. Because we are dealing with an infinitely small pyramid, aside from quantities of a higher order of smallness compared to S_ν, we can write

$$\mathbf{F}_1 = -\boldsymbol{\sigma}_1 S_1, \qquad \mathbf{F}_2 = -\boldsymbol{\sigma}_2 S_2, \qquad \mathbf{F}_3 = -\boldsymbol{\sigma}_3 S_3, \qquad \mathbf{F}_\nu = \boldsymbol{\sigma}_\nu S_\nu.$$

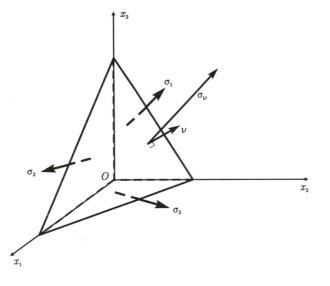

Figure 2.13 Triangular pyramid inside a material, showing reaction stresses that act over the various faces.

On the same pyramid, there also act volume-distributed forces (of the same type as gravity) that are proportional to the pyramid volume V: we write $\mathbf{W} = \mathbf{w}V$ for these (see Figure 2.14). We again use the solidification principle: the pyramid is in equilibrium when deformed, and so these forces behave exactly as for a rigid pyramid. By classical mechanics, the vector sum of all forces should be zero in equilibrium:

$$\mathbf{F}_1 + \mathbf{F}_2 + \mathbf{F}_3 + \mathbf{F}_\nu + \mathbf{W} = \mathbf{0}.$$

Substituting the above relations into this, we get

$$-\boldsymbol{\sigma}_1 S_1 - \boldsymbol{\sigma}_2 S_2 - \boldsymbol{\sigma}_3 S_3 + \boldsymbol{\sigma}_\nu S_\nu + \mathbf{w}V + \left(\begin{array}{c}\text{higher-order}\\\text{terms}\end{array}\right) = \mathbf{0}.$$

Here the "higher-order terms" are terms of a higher order of smallness compared to S_ν. Now let us divide through by S_ν:

$$-\boldsymbol{\sigma}_1 \frac{S_1}{S_\nu} - \boldsymbol{\sigma}_2 \frac{S_2}{S_\nu} - \boldsymbol{\sigma}_3 \frac{S_3}{S_\nu} + \boldsymbol{\sigma}_\nu + \mathbf{w}\frac{V}{S_\nu} + \frac{\left(\begin{array}{c}\text{higher-order}\\\text{terms}\end{array}\right)}{S_\nu} = \mathbf{0}.$$

The coordinate faces of the pyramid are projections of the inclined face onto the corresponding coordinate planes; hence, the areas relate through the formula

$$S_k = S_\nu \cos(\widehat{\nu, \mathbf{i_k}}) = S_\nu n_k.$$

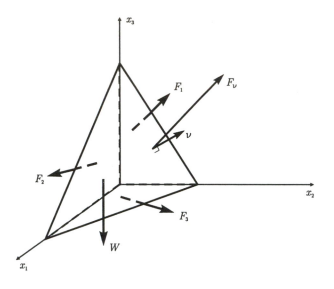

Figure 2.14 Forces acting on the elementary pyramid.

So, the formula can be rewritten as

$$-\sigma_1 n_1 - \sigma_2 n_2 - \sigma_3 n_3 + \sigma_\nu + \mathbf{w}\frac{V}{S_\nu} + \frac{\left(\begin{array}{c}\text{higher-order}\\ \text{terms}\end{array}\right)}{S_\nu} = \mathbf{0}.$$

Now, let us perform a limit passage under which $S_\nu \to 0$, so that the pyramid shrinks while maintaining its shape and orientation. During this passage, the first four terms tend to $-\sigma_1 n_1 - \sigma_2 n_2 - \sigma_3 n_3 + \sigma_\nu$, taken at a point corresponding to the infinitesimal area with normal ν, whereas the last two terms tend to zero (indeed, V/S_ν is one-third the height of the pyramid that tends to zero, and the last term tends to zero by definition of smallness). Thus, we get exactly the equation (2.16) that we wanted to derive.

Note that the inertia forces of a small portion are proportional to the volume of the portion; this means they have the same order of smallness as the force $\mathbf{w}V$, and thus (2.16) *holds not only for the equilibrium state of a body, but for dynamical states as well.*

Let us rewrite the vector equation (2.16) in component form:

$$\sigma_{\nu 1} = \sigma_{11} n_1 + \sigma_{21} n_2 + \sigma_{31} n_3,$$
$$\sigma_{\nu 2} = \sigma_{12} n_1 + \sigma_{22} n_2 + \sigma_{32} n_3,$$
$$\sigma_{\nu 3} = \sigma_{13} n_1 + \sigma_{23} n_2 + \sigma_{33} n_3. \tag{2.17}$$

This relation is important. As a transformation law, it differs from that under which vector components transform with changes of coordinate frame, and this means that in stress we have encountered an object whose nature is not vectorial. An object in three-dimensional space that takes on the appearance of a vector over any elementary area, but whose components on an arbitrary area are related by a rule having the form of (2.17), is called a *tensor of the second rank.*

A tensor is a mathematical object, a quantity, characterized by components connected to the frame directions and to scaling of the coordinate axes. Like a vector, a second-rank tensor is uniquely determined by its principal components. Hence, we will talk about the *stress tensor* and denote the collection of its components by $\{\sigma_{ij}\} = \{\sigma_{11}, \sigma_{12}, \ldots, \sigma_{33}\}$. In any other frame, these can be recalculated by formulas of the type (2.17) (or by more general but similar ones for non-Cartesian frames).

Let us return to the stress tensor $\{\sigma_{ij}\}$ (observe that a certain Cartesian frame is implicit in this notation). In the theory of elasticity, this tensor is presented as a matrix,

$$\begin{pmatrix} \sigma_{11} & \sigma_{12} & \sigma_{13} \\ \sigma_{21} & \sigma_{22} & \sigma_{23} \\ \sigma_{31} & \sigma_{32} & \sigma_{33} \end{pmatrix},$$

and is usually denoted by a single letter, say $\boldsymbol{\sigma}$. Its second-rank tensor nature can be emphasized by writing the symbol $\underline{\underline{\sigma}}$ instead, which is done in handwriting.

Considering the equilibrium of a solid body in classical mechanics, we saw that we must have both the resultant force and the resultant moment equal to zero. The moment equation gives us additional properties of the stress tensor. However, consideration of this for the pyramid brings unnecessary complications. So Cauchy proposed that we consider the equilibrium of an infinitely small cube whose faces parallel the corresponding frame axes. We will not derive the moment equation here — the derivation is simple but cumbersome. In the limit passage, the force equation is automatically satisfied. The moment equation yields a property known as the *symmetry* of the stress tensor, expressed by the three relations

$$\sigma_{12} = \sigma_{21}, \qquad \sigma_{13} = \sigma_{31}, \qquad \sigma_{23} = \sigma_{32}.$$

Thus, the stress tensor at a point is uniquely determined by only six of its components.

We began by saying that we would consider a "deformed" portion of the material. Such a statement is to be found in any textbook, and few students bother to question it.[6] But imagine that we could "paint" a small pyramid in the deformed body, and consider what its shape might have been in the nondeformed state. The faces would become skewed (possibly non-planar), the face areas and angles would change, and so on. This means that the average value of stress, recalculated for the undeformed state of the material,

---

[6]Seemingly innocuous statements like this occur in many branches of mathematics. Sometimes, however, they turn out to be crucial assumptions having vast consequences. Such is the case here.

would change a bit. When we consider infinitely small deformations, we get the same quantities in the limit passage; however, if we begin to consider finite deformations, then we can get significant differences in all values of the recalculated stresses. This brings us into the field of nonlinearity. Nonlinear models describe nature more accurately, but at the price of more complex reasoning and calculation; in the presence of nonlinearity, we cannot apply superposition. Many nonlinear problems still represent open questions in continuum mechanics, and the interested reader can explore thick volumes on the subject.

In linear elasticity, we do not distinguish between the stress tensors relating to the deformed or the nondeformed states; we use only the nondeformed state as our reference. But in nonlinear theories, we encounter nonunique forms of the stress tensor that depend on the reference state. The tensor we have introduced above is called the *Cauchy stress tensor*. Its components are defined for the reference frame in the deformed state.

2.8 Principal Axes and Invariants of the Stress Tensor

We have derived formula (2.16) for a stress vector on an elementary area with unit normal $\boldsymbol{\nu} = (n_1, n_2, n_3)$:

$$\boldsymbol{\sigma}_{\nu} = \boldsymbol{\sigma}_1 n_1 + \boldsymbol{\sigma}_2 n_2 + \boldsymbol{\sigma}_3 n_3.$$

Let us consider a question. At a given point, is it possible to orient an area element in such a way that no shear stress acts on the element? A purely normal stress would be similar to the pressure acting at a point in a liquid. Let us set $\boldsymbol{\sigma}_{\nu} = p\boldsymbol{\nu} = (pn_1, pn_2, pn_3)$, and write the above equation in component form. After a bit of rearrangement, we have

$$\sigma_{11} n_1 + \sigma_{21} n_2 + \sigma_{31} n_3 - p n_1 = 0,$$
$$\sigma_{12} n_1 + \sigma_{22} n_2 + \sigma_{32} n_3 - p n_2 = 0,$$
$$\sigma_{13} n_1 + \sigma_{23} n_2 + \sigma_{33} n_3 - p n_3 = 0. \qquad (2.18)$$

These can be regarded as three simultaneous linear equations in the three unknowns n_1, n_2, n_3, and the system has a nonzero solution if and only if its determinant is equal to zero:

$$\begin{vmatrix} \sigma_{11} - p & \sigma_{12} & \sigma_{13} \\ \sigma_{21} & \sigma_{22} - p & \sigma_{23} \\ \sigma_{31} & \sigma_{32} & \sigma_{33} - p \end{vmatrix} = 0. \qquad (2.19)$$

With respect to p, this is the third-order algebraic equation

$$p^3 - I_1 p^2 + I_2 p - I_3 = 0, \qquad (2.20)$$

where

$$I_1 = \sigma_{11} + \sigma_{22} + \sigma_{33},$$
$$I_2 = \sigma_{11}\sigma_{22} + \sigma_{22}\sigma_{33} + \sigma_{33}\sigma_{11} - \sigma_{12}{}^2 - \sigma_{23}{}^2 - \sigma_{31}{}^2,$$
$$I_3 = \sigma_{11}\sigma_{22}\sigma_{33} + 2\sigma_{12}\sigma_{23}\sigma_{31} - \sigma_{11}\sigma_{23}{}^2 - \sigma_{22}\sigma_{31}{}^2 - \sigma_{33}\sigma_{12}{}^2.$$

This is the *characteristic equation* for the tensor σ. It has no more than three different roots, at least one of which should be real. However, introducing the matrix $A = (\sigma_{ij})$, we see that (2.18) can be rewritten in matrix form as

$$A\nu = p\nu,$$

and thus our problem involves finding the eigenvalues p_k and eigenvectors ν_k (nontrivial vector solutions of this are defined up to the magnitude, so we should add the equation $|\nu| = 1$ and show which of the two directions should be taken). For a symmetric matrix, it is known that the eigenvalues are real, and that eigenvectors corresponding to different eigenvalues are orthogonal. This means that if there are three different eigenvalues p_k, then there are three mutually orthogonal directions representing normals to area elements over which the vector shear stresses vanish. If we get two equal eigenvalues, then there is a plane with the property that the vector stress is purely normal for any area element whose normal lies in this plane. If we get three equal eigenvalues, then the stress vector is normal to all area elements through a given point and is equal in all directions, so that is similar to the case of an ideal frictionless liquid.

In any case, at a given point, we can find three mutually orthogonal directions for which the stress is orthogonal to the area element. These are the *principal directions* of the stress tensor at the point, and the corresponding p_k are the *principal stresses*.

Using the principal stresses and the theorem of Francois Viète (1540–1603) for the cubic equation, we have

$$I_1 = p_1 + p_2 + p_3,$$
$$I_2 = p_1p_2 + p_2p_3 + p_3p_1,$$
$$I_3 = p_1p_2p_3.$$

These three values are the *principal invariants* of the stress tensor. They do not depend on the orientation of the coordinate frame, and the same can be said of the p_k: these are objective characteristics of the state of stress in a body at a point.

The principal stresses and directions give us a picture of the state of stress at a point. Because they change from point to point, however, it is difficult to make use of this picture during the analytical solution of a problem.

Note that we exploited only the symmetry of the stress tensor in this development. Hence, we could introduce similar notions and terminology (characteristic equation, principal directions, etc.) for any symmetric tensor of the second rank. We shall do this for the strain tensor, and no additional explanation should be required.

2.9 On the Continuum Model and Limit Passages

The models of continuum mechanics reflect everyday experience. They assume we can ignore the atomic structure of materials, and hence are inherently approximate in nature. A continuum model cannot predict what happens inside a body composed of ten atoms; it can only predict macroscopic characteristics of deformations in which large ensembles of atoms are involved. Use of the theory is justified on the grounds that everyday items are composed of a great many atoms. In washing our hands with soap, for instance, we use more than 10^{15} soap molecules, and we cannot distinguish one molecule from another in this process.

The approach in which a medium is regarded as continuous — without gaps — is popular in sciences that study large numbers of similar things. In population biology, one can consider the distribution of fish in an ocean as something continuous, and when we ask how many exist per unit volume, we may get an answer like 2.5. This would, of course, represent an average figure: a certain small volume at a given time may actually contain 10 fish or none at all, but if we were to count the fish in a cubic mile of water, we would find the average value to hold well (within 5%, say).

But the continuum model has obvious defects. One of its strict consequences shows up in hydromechanics: a surface that at one time consists of material points of the medium contains the same material points and remains a surface for all time. In particular, this means that the points inside any closed material surface must remain inside that surface during whatever deformations may occur. Although this model is applied to gases as well as fluids, common sense tells us that the molecules of a gas can move freely back and forth across any surface without being restricted in this way. Only in an average sense can we assert that inside some material surface the amount of material stays constant.

The continuity hypothesis is firm and allows us to use the tools of calculus such as limit passages, derivatives, and integrals. These procedures could be almost senseless if applied at a point inside a material having atomic structure. In older books on continuum mechanics, we find attempts at justification of limit passages such as the one we performed for the stress tensor in the previous section. We are instructed to consider smaller and smaller elementary areas or volumes until several hundred atoms or molecules are involved. The idea was that the elementary area or volume under consideration would then be "small," while the number of particles involved would still be "large." Unfortunately, such reasoning is fallacious. We cannot interrupt any limit passage and stop at a finite value, regardless of what our intuition tells us is "small enough." We have said that the number $100^{-100000}$ is not infinitely small: it will be small in some circumstances and large in others, depending on what one is measuring. So the proper procedure is to introduce a truly continuous model of a medium, forgetting completely about atoms and molecules; then, after obtaining some result from the model, we can discuss the range of applicability of the model to various amounts of the material. We should mention that some atomic theories, like the theory of cracks, use the results of continuum mechanics as intermediate results in situations when even just a few atoms are under consideration.

Our next question is as follows. Given the well-established atomic structure of materials and the power of modern computers, would it be more reasonable to describe the behavior of materials from the atomic point of view? Should we attempt to write down the equations of motion for each atom and let a fast computer handle the details?

The answer is no. The real atomic structure of materials is so complex that we cannot describe it for large material samples. Materials we normally regard as homogeneous are actually quite inhomogeneous. Even water possesses additional structure beyond the familiar H_2O structure of its molecules. Steel has domains, a kind of imperfect grain structure inside the material. Indeed, any real material will be imperfect in some sense, and the distribution of molecules inside any body will be unknown as a result. Through painstaking effort we are sometimes able to produce materials that are more or less homogeneous. We find that such materials are much stronger than ordinary materials, and have other interesting characteristics. The continuum hypothesis serves to regularize the structure of a material, and therefore allows us to predict many material characteristics and processes in a relatively simple fashion. By contrast, an exact description could yield worse results because of the prohibitive number of calculations that would have to be performed (recall our previous discussion of roundoff error in numerical calculations).

Thus, the continuum model yields solutions to many problems of ordinary life. This being said, continuum mechanicists will continue to seek new ways to represent the atomic properties of materials. The approximate nature of the continuum model suggests that we might be able to use discrete models for the same processes. Such models now exist under the names finite element method and boundary element method. The equations for these discrete models are based on the continuum model, but some problems are treated by direct consideration of "small but finite" volumes, or other characteristics. Discrete models are often judged to be good because various limit passages applied to them yield the same differential equations as would be obtained using the calculus of differentiable functions. But such models can survive on their own as fully independent approximations, and the only real judge of a model is comparison of its predictions with the results of experiment.

We might finish this section with some further remarks about limit passages and the order in which they are performed. It is well known in calculus that a multiple limit passage can give different results when done in different orders: for example, the derivative of an integral is not always equal to the integral of the derivative. This situation arises in any physical science: first we use limit passages to derive the governing equation for a particular model; then we perform various other limit passages, substitutions, and so forth, on the model, in order to apply it. We could ask whether the result would be the same if we were to begin with small quantities, particles, and so on, produce all the operations in discrete form; and then perform all necessary limit passages at once. Nobody has really tried to analyze problems from this viewpoint, and the situation is so complex that we do not know the circumstances under which the result would be correct. Our experience with multiple limit theorems in calculus suggests that if all the results are smooth enough, then everything should be fine. But this is only a feeling. Stronger validation comes from either direct or indirect experiment. Only when experimental results differ from our predictions do we think that something must be wrong. As a rule, this leads us to other models of the same processes — models that can fit our experimental data to within needed accuracy.

2.10 Equilibrium Equations

At this point, it is convenient to formulate the equilibrium equations for a continuum body. We recall that these do not depend on the constitutive equation that relates stresses to strains within the body.

When considering the equilibrium problems for a wire and a ruler, we considered a small part of the body in question. We shall do the same here. Before proceeding, we recall a simple fact. For a continuously differentiable function $f(x)$ in one variable, we have

$$f(x + \Delta x) = f(x) + f'(x)\Delta x + \omega(x, \Delta x), \qquad (2.21)$$

where the function $\omega = \omega(x, \Delta x)$ is such that $\lim_{\Delta x \to 0} \omega(x, \Delta x)/\Delta x = 0$. The term $f'(x)\Delta x$ is the main (linear) part of the increment of $f(x)$ for the increment Δx.

For a vector function

$$\mathbf{f}(x) = (f_1(x), f_2(x), f_3(x)),$$

we introduce the derivative with respect to x as

$$\mathbf{f}'(x) = (f'_1(x), f'_2(x), f'_3(x)).$$

This formula comes from the general definition of the derivative

$$\mathbf{f}'(x) = \lim_{\Delta x \to 0} \frac{\mathbf{f}(x + \Delta x) - \mathbf{f}(x)}{\Delta x}.$$

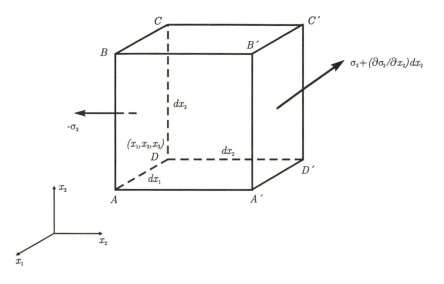

Figure 2.15 Opposing stresses acting on two sides of an elementary parallelepiped.

Formula (2.21) for the increments of the components of a vector function becomes

$$\mathbf{f}(x + \Delta x) = \mathbf{f}(x) + \mathbf{f}'(x)\Delta x + \boldsymbol{\omega}(x, \Delta x), \tag{2.22}$$

where the components of $\boldsymbol{\omega}$ satisfy $\lim_{\Delta x \to 0} \omega_i(x, \Delta x)/\Delta x = 0$.

Now let us return to the equilibrium equations. Consider an infinitesimal rectangular parallelepiped, with a vertex (x_1, x_2, x_3) and sides dx_1, dx_2, dx_3, in a deformed three-dimensional body, as shown in Figure 2.15. Note that in introducing the stress tensor we were interested only in the components of the tensor, so we used quantities of the same order of magnitude as the stresses, which should be finite, and, during limit passages, all higher-order terms disappeared. Now, because we seek to better describe the equilibrium of a small portion of deformed material, we retain terms that have the same order of smallness as the sizes of the parallelepiped (and hence we suppose that the stress components are continuously differentiable functions). In this way, we derive the equilibrium equation for an infinitely small parallelepiped under external load and under the reactions of the rest of the body.

In Figure 2.15, we denote by $-\boldsymbol{\sigma}_2$ the vector of averaged stresses acting on face $ABCD$ and directed outward from the parallelepiped. The average stress on a face is obtained as the ratio of the corresponding resultant reaction force over the face to the area of the face, and so it is a bit different from $-\boldsymbol{\sigma}_2$ but tends to $-\boldsymbol{\sigma}_2$ when the face area tends to zero. The resultant force on a domain is found by integrating the stress vector over that domain. The minus sign in $-\boldsymbol{\sigma}_2$ is present because the outward normal to the face has the direction opposite to that of the x_2-axis. Recall that on the opposite side $A'B'C'D'$ of the parallelepiped, the vector of averaged stresses changes a

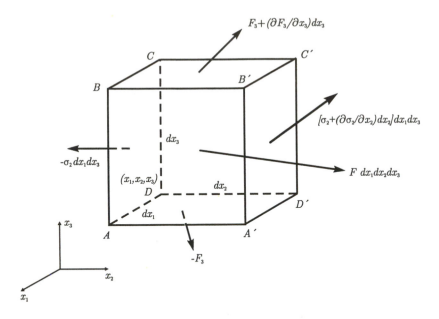

Figure 2.16 Forces acting on an elementary parallelepiped.

little because of the other forces acting on the parallelepiped. Using formula (2.22), we have on face $A'B'C'D'$,

$$\sigma_2(x_2 + dx_2) = \sigma_2(x_2) + \sigma'_2(x_2)\, dx_2 + \left(\begin{array}{c} \text{term of higher order} \\ \text{of smallness than } dx_2 \end{array} \right).$$

On the two other pairs of opposing sides of the parallelepiped, we have similar formulas for the averaged stresses:

$$\sigma_1(x_1 + dx_1) = \sigma_1(x_1) + \sigma'_1(x_1)\, dx_1 + \left(\begin{array}{c} \text{term of higher order} \\ \text{of smallness than } dx_1 \end{array} \right),$$

$$\sigma_3(x_3 + dx_3) = \sigma_3(x_3) + \sigma'_3(x_3)\, dx_3 + \left(\begin{array}{c} \text{term of higher order} \\ \text{of smallness than } dx_3 \end{array} \right).$$

Here, we have shown in the σ_i only an argument that changes in the calculation; however, each function depends, in fact, on all three arguments x_1, x_2, x_3 and the areas of the faces.

Now let us write out the equilibrium equation for this "solidified" parallelepiped. The resultant vector force on a face is an "averaged vector stress" multiplied by the face area. For example, on face $ABCD$, it is $-\sigma_2(x_2)\, dx_1\, dx_3$; in Figure 2.16, we show all forces acting on the parallelepiped. Here, $\mathbf{F}$ is the density of volume forces. We use the notation $\mathbf{F}_1$

for $\boldsymbol{\sigma}_1(x_1)\, dx_2\, dx_3$ and $\mathbf{F}_3$ for $\boldsymbol{\sigma}_3(x_3)\, dx_1\, dx_2$:

$$
\begin{aligned}
&(\boldsymbol{\sigma}_2(x_2 + dx_2)\, dx_1\, dx_3 - \boldsymbol{\sigma}_2(x_2)\, dx_1\, dx_3) \\
&+ (\boldsymbol{\sigma}_1(x_1 + dx_1)\, dx_2\, dx_3 - \boldsymbol{\sigma}_1(x_1)\, dx_2\, dx_3) \\
&+ (\boldsymbol{\sigma}_3(x_3 + dx_3)\, dx_1\, dx_2 - \boldsymbol{\sigma}_3(x_3)\, dx_1\, dx_2) \\
&+ \mathbf{F}\, dx_1\, dx_2\, dx_3 = \mathbf{0}.
\end{aligned}
$$

Substituting the above formulas for the increments, we have

$$
\begin{aligned}
&(\boldsymbol{\sigma}_2(x_2)\, dx_1\, dx_3 + \boldsymbol{\sigma}'_2(x_2)\, dx_2\, dx_1\, dx_3 - \boldsymbol{\sigma}_2(x_2)\, dx_1\, dx_3) \\
&+ (\boldsymbol{\sigma}_1(x_1)\, dx_2\, dx_3 + \boldsymbol{\sigma}'_1(x_1)\, dx_1\, dx_2\, dx_3 - \boldsymbol{\sigma}_1(x_1)\, dx_2\, dx_3) \\
&+ (\boldsymbol{\sigma}_3(x_3)\, dx_1\, dx_2 + \boldsymbol{\sigma}'_3(x_3)\, dx_3\, dx_1\, dx_2 - \boldsymbol{\sigma}_1(x_1)\, dx_1\, dx_2) \\
&+ \mathbf{F}\, dx_1\, dx_2\, dx_3 + \left(\begin{array}{c} \text{terms of higher smallness} \\ \text{than } dx_1 dx_2 dx_3 \end{array} \right) = \mathbf{0}.
\end{aligned}
$$

Dividing through by $dx_1\, dx_2\, dx_3$ and letting all dimensions of the elementary parallelepiped tend to zero proportionally, we get

$$
\frac{\partial \boldsymbol{\sigma}_1}{\partial x_1} + \frac{\partial \boldsymbol{\sigma}_2}{\partial x_2} + \frac{\partial \boldsymbol{\sigma}_3}{\partial x_3} + \mathbf{F} = \mathbf{0}. \tag{2.23}
$$

This limit passage, provided that the stress functions are sufficiently smooth, yields the stress values $\boldsymbol{\sigma}_k(x_1, x_2, x_3)$ of the previous section. While the averaged stresses were functions in one variable, the stresses themselves are functions in three variables. The reader is aware that for such functions we can introduce differentiation with respect to any one of these variables, fixing the other two. In this way, we have three different derivatives with respect to the x_i ($i = 1, 2, 3$). The expressions for these depend on the other two variables. They are called *partial derivatives* and are denoted by $\partial f / \partial x_i$. The reader may review this with the help of any calculus book if necessary. Of more interest right now is that such derivative-type limit passages of averaged functions and their correspondences to partial derivatives are considered evident in mechanics, but are not easily justified mathematically — especially if one seeks minimal conditions under which they are valid. Mechanics textbooks seldom focus on such issues, assuming instead that all functions involved are as smooth as necessary. Upon completing their derivations under such assumptions, these textbooks proceed to apply the results to situations in which those same assumptions are violated at certain points, or even over entire surface or volume regions. But only a disagreement with experimental results will in general cause mechanicists to reconsider their assumptions.

We have mentioned that the stresses $\boldsymbol{\sigma}_k$ are made up of components of the stress tensor. As such they are not vectors, because they are not invariant under changes of frame; however, our use of only one coordinate frame has allowed us to use them as if they were real vectors. Of course, in a rotated

Cartesian coordinate system, the vector equilibrium equation would have the same form, but the values σ_k cannot be transformed using the formulas for vector transformation. Instead, we need tensor transformation rules. However, this trick of using fictitious vectors is convenient and common in mechanics derivations.

Now, let us rewrite the equation in component form:

$$\frac{\partial \sigma_{11}}{\partial x_1} + \frac{\partial \sigma_{21}}{\partial x_2} + \frac{\partial \sigma_{31}}{\partial x_3} + F_1 = 0,$$

$$\frac{\partial \sigma_{12}}{\partial x_1} + \frac{\partial \sigma_{22}}{\partial x_2} + \frac{\partial \sigma_{32}}{\partial x_3} + F_2 = 0,$$

$$\frac{\partial \sigma_{13}}{\partial x_1} + \frac{\partial \sigma_{23}}{\partial x_2} + \frac{\partial \sigma_{33}}{\partial x_3} + F_3 = 0. \tag{2.24}$$

This concludes our efforts in this section. We emphasize only that these follow from the equilibrium equations for an elementary parallelepiped, taking into account some higher-order force terms. What happens if we include even higher order stress terms? For the model we consider, this does not bring in other equations, although this is not true for all models.

This system consists of three equations in six independent variables (six because of the symmetry; of the nine components of $\boldsymbol{\sigma}$, only six are independent). Experience tells us that, normally, for well-posedness of a problem,[7] the number of equations should equal the number of unknowns. This rule is not absolute, but it does hold in the present case: we need more than just equilibrium equations to describe what happens to a body under load. We did not use the physical properties of the body, and our experience indicates that the resulting state can be quite different, depending on the material properties. For a proper setup of an equilibrium problem, we should supplement the equilibrium equations with a constitutive law for the material, and for this, the notion of strains is required.

2.11 The Strain Tensor

Now we consider another main constituent of the analysis of deformation for a three-dimensional body: the strain tensor. For a wire, we showed that the strain ε at a point is given by $\varepsilon = du/dx$, where $u = u(x)$ is the function of displacement of a point x of the wire. We will generalize this to three dimensions.

At a point $\mathbf{x} = (x_1, x_2, x_3)$ in a body, we introduce the vector of displacement,

$$\mathbf{u} = \mathbf{u}(\mathbf{x}) = (u_1(\mathbf{x}), u_2(\mathbf{x}), u_3(\mathbf{x})),$$

---

[7]In mechanics, well-posedness normally means that a problem should have a unique solution that depends continuously on a change in parameters of the problem. This is typical for linear problems. The uniqueness requirement can be omitted for nonlinear problems.

of the point after deformation. For the arguments of functions, we will interchangeably use vector and scalar notations: $f(\mathbf{x}) = f(x_1, x_2, x_3)$.

Now let us consider how to describe small deformations of a straight segment $dx = (dx_1, dx_2, dx_3)$ in the body. As usual in mechanics, we use the notation of the differential for dx, as though this segment were infinitely small, but we shall at first treat it as a nonzero finite quantity. The displacement of a point $\mathbf{x}$ after deformation is $\mathbf{u}(\mathbf{x})$, whereas the displacement of the other end of the small segment dx (with reference vector $\mathbf{x} + dx$) is $\mathbf{u}(\mathbf{x} + dx)$. Thus, the change in the vector connecting the new endpoints of the deformed segment is $\mathbf{u}(\mathbf{x} + dx) - \mathbf{u}(\mathbf{x})$. This is the increment of the function $\mathbf{u}(\mathbf{x})$ corresponding to the increment of the argument $\mathbf{x}$.

We suppose that $\mathbf{u}(\mathbf{x})$ is sufficiently smooth. For a steel wire, we know that the strain $\varepsilon = du/dx$, which is dimensionless, cannot normally exceed 0.001 or so, and this quantity is considered as "infinitely small" in elasticity theory. Thus, quantities like ε^2 can be neglected in comparison with ε.

We can derive the strain for each direction in a three-dimensional body in the same way that we derived the strain for a wire. This means that the quantity $\partial u_k/\partial x_k$ is the strain for an elementary segment parallel to the x_k-axis. We suppose all the strains to be "infinitely small" (in the above sense). However, there remain other partial derivatives of u_i. It is clear that, besides the deformation, $\partial u_i/\partial x_j$ characterizes the rotation of an elementary segment dx. We will suppose this is also infinitely small, and so all the $\partial u_i/\partial x_j$ must be infinitely small in the above sense. So, when we meet a sum of a partial derivative of u_i and a squared partial derivative, we will neglect the latter.

We have used the fundamental increment formula to find a linear approximation for a function in one variable. A similar formula is used for a function in several variables:

$$u(\mathbf{x} + dx) = u(\mathbf{x}) + du(\mathbf{x}, dx) + \text{higher-order terms in } |dx|$$

$$= u(\mathbf{x}) + \frac{\partial u}{\partial x_1} dx_1 + \frac{\partial u}{\partial x_2} dx_2 + \frac{\partial u}{\partial x_3} dx_3 + \text{higher-order terms.}$$

We will use this for each component of the vector function $\mathbf{u}(\mathbf{x})$. After deformation, the point $\mathbf{x}$ has relocated to the point $\mathbf{x} + \mathbf{u}$, whereas the point $\mathbf{x} + dx$ after deformation becomes $\mathbf{x} + dx + \mathbf{u} + du$ up to terms of the second order of smallness with respect to $|dx|$. The corresponding segment connecting these points is, by the last equation,

$$dx + d\mathbf{u} = \left(dx_1 + \frac{\partial u_1}{\partial x_1} dx_1 + \frac{\partial u_1}{\partial x_2} dx_2 + \frac{\partial u_1}{\partial x_3} dx_3, \right.$$
$$dx_2 + \frac{\partial u_2}{\partial x_1} dx_1 + \frac{\partial u_2}{\partial x_2} dx_2 + \frac{\partial u_2}{\partial x_3} dx_3,$$
$$\left. dx_3 + \frac{\partial u_3}{\partial x_1} dx_1 + \frac{\partial u_3}{\partial x_2} dx_2 + \frac{\partial u_3}{\partial x_3} dx_3 \right). \qquad (2.25)$$

First, we would like to find the strains along any direction at a point. A strain is defined as the ratio of the increment in length of a segment after deformation to its length before deformation. (Note that even here we used the smallness of a deformation: being deformed, a straight segment could be bent. But under our assumptions, we neglect quantities of a higher order of smallness, like the deflection. We do this even in the basic terms we use.) Finding the magnitude of a vector quantity involves nonlinear operations like squaring and taking square roots. Because we must extract the linear part from the function increment, we would prefer to avoid squaring variables. This can be done in a simple fashion if we see that, for a wire, the strain can be obtained from the identity

$$\frac{\Delta L}{L} = \frac{(L + \Delta L)^2 - L^2}{2L^2} - \frac{\Delta L^2}{2L^2},$$

which is linear in ΔL. We can apply the same tool to any elementary segment defined by a vector (dx_1, dx_2, dx_3) in the material, considering square expressions for $|d\mathbf{x}|$ and the deformed segment defined by the expression $|d\mathbf{x} + d\mathbf{u}|$, up to terms of the second order of smallness. We keep in mind that at this stage we regard the dx_i as small but finite, whereas all the first partial derivatives of u_i are so small that we will neglect squared derivatives in expressions containing linear terms (i.e., linear with respect to the derivatives). The above example for the wire shows that the linear part in $\partial u_i / \partial x_j$ of the expression

$$\frac{(d\mathbf{x} + d\mathbf{u})^2 - d\mathbf{x}^2}{2\,d\mathbf{x}^2} \tag{2.26}$$

gives us the strain in the direction $d\mathbf{x}$.

Bringing in equation (2.25), we get in the numerator (divided by 2) the following expression:

$$\frac{\partial u_1}{\partial x_1}\,dx_1\,dx_1 + \frac{\partial u_2}{\partial x_2}\,dx_2\,dx_2 + \frac{\partial u_3}{\partial x_3}\,dx_3\,dx_3 + \left(\frac{\partial u_1}{\partial x_2} + \frac{\partial u_2}{\partial x_1}\right)dx_1\,dx_2$$

$$+ \left(\frac{\partial u_1}{\partial x_3} + \frac{\partial u_3}{\partial x_1}\right)dx_1\,dx_3 + \left(\frac{\partial u_2}{\partial x_3} + \frac{\partial u_3}{\partial x_2}\right)dx_2\,dx_3 + \left(\begin{array}{c}\text{higher-order}\\\text{terms}\end{array}\right).$$

Using this, we see that if we take a small segment $(dx_1, 0, 0)$, the main part of (2.26) is $\varepsilon_{11} = \partial u_1 / \partial x_1$, which is the strain in the x_1-direction, as we have said. (Of course, here we should produce the limit passage. So at this point, we consider dx_1 to be infinitely small, hence it becomes a differential.) In a similar way the strains in the x_2- and x_3-directions are $\varepsilon_{22} = \partial u_2 / \partial x_2$ and $\varepsilon_{33} = \partial u_3 / \partial x_3$, respectively. However, in the above expression there remain other terms. Let us consider their meanings. Let the displacement vector have the form $\mathbf{u} = (\gamma x_2, 0, 0)$. If we take a unit cube, after the transformation it would be "skewed" along the x_1-direction, as shown in Figure 2.17, and thus for small shears, the tangent γ of the shear angle is

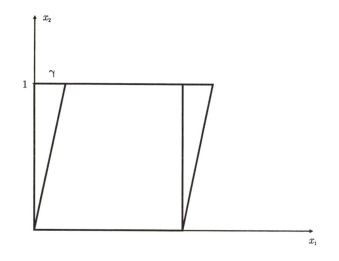

Figure 2.17 Geometrical meaning of the mixed components of the strain tensor:
for small γ, we have $\varepsilon_{12} \approx \gamma/2$.

approximately equal to the angle itself. The only nonzero component for this
transformation is $\partial u_1/\partial x_2 + \partial u_2/\partial x_1 = \gamma$, so this term has meaning as the
shear angle. However, we introduce a notation for one-half of this quantity:
$\varepsilon_{12} = \frac{1}{2}(\partial u_1/\partial x_2 + \partial u_2/\partial x_1) = \gamma/2$. For the sake of symmetry, we then
introduce $\varepsilon_{21} = \varepsilon_{12}$. In a similar manner, we can introduce other "mixed"
components $\varepsilon_{13} = \varepsilon_{31}$ and $\varepsilon_{23} = \varepsilon_{32}$. In this way, we have introduced a
matrix of strain coefficients

$$\begin{pmatrix} \varepsilon_{11} & \varepsilon_{12} & \varepsilon_{13} \\ \varepsilon_{21} & \varepsilon_{22} & \varepsilon_{23} \\ \varepsilon_{31} & \varepsilon_{32} & \varepsilon_{33} \end{pmatrix} \qquad (2.27)$$

with

$$\varepsilon_{ij} = \frac{1}{2}\left(\frac{\partial u_i}{\partial x_j} + \frac{\partial u_j}{\partial x_i}\right).$$

These are the components of an object called the tensor of small strains
(or simply, *strain tensor*). In old-fashioned books, one can find double quan-
tities for the mixed components $\varepsilon_{12}, \varepsilon_{13}, \varepsilon_{23}$. This is to keep the geometrical
meaning of these components as shear angles. However, only in our notation
is this object a symmetric tensor of the second rank. The latter means that
under frame transformations its components change according to the same
rules as for other second-rank tensors such as the stress tensor or inertia
tensor. It is also an objective quantity, which means that if we find its components in a frame, then they are equal to the ones recalculated from any
other frame by the general rules for the transformation of tensor components.

The fact that the strains ε_{ij} constitute a symmetric tensor of the second rank[8] brings us new results immediately. At any point, there are three mutually orthogonal principal directions along which shear strains are absent and only principal strains e_1, e_2, e_3 exist. The deformations in these directions are of the stretching type only. If we take an elementary cube

$$dx_1 = dx_2 = dx_3$$

with respective faces parallel to the principal directions, then after deformation it becomes a rectangular parallelepiped with sides

$$dx_1(1 + e_1), \qquad dx_2(1 + e_2), \qquad dx_3(1 + e_3).$$

The volumes of the cube and the corresponding parallelepiped are

$$V_0 = dx_1\, dx_2\, dx_3, \qquad V_1 = dx_1(1 + e_1)\, dx_2(1 + e_2)\, dx_3(1 + e_3).$$

Thus, keeping in mind that we consider the tensor of small deformation so that all higher powers of ε_{ij} are neglected in comparison with linear terms, the change in the volume due to deformation is

$$\Delta V = (e_1 + e_2 + e_3)\, dx_1\, dx_2\, dx_3.$$

The relative change in volume is

$$\theta = \frac{\Delta V}{V_0} = e_1 + e_2 + e_3.$$

This is the first invariant for the strain tensor, and it measures the relative change of volume not only for a cube but for any elementary volume containing the point of interest. It can be expressed in any Cartesian frame as

$$\theta = \varepsilon_{11} + \varepsilon_{22} + \varepsilon_{33}.$$

For an incompressible body (i.e., a body for which any volume after deformation is equal to the volume before deformation) we have the equation of incompressibility

$$\varepsilon_{11} + \varepsilon_{22} + \varepsilon_{33} = 0. \tag{2.28}$$

---

[8]This fact must be verified, not simply stated. For this, we should show that components of strain in any direction transform under the same rules as the components of the stress tensor that we derived earlier. The reader can try to verify this him- or herself, or consult any book on the theory of elasticity.

2.12 Generalized Hooke's Law

To complete the set of equations needed to describe the deformation of an elastic body, we must derive an analogue of Hooke's law for the three-dimensional case. It is clear that this should link the stress and strain tensors. These tensors are not simply proportional; this is shown by the simple fact that a rubber band becomes narrower when stretched. The same holds for any material: when we stretch a steel rod, its width dimensions decrease slightly. These deformations are tiny and require precise instruments for their measurement, although if we hang a weight from a long, thin wire, we can actually see the stretching effect. The reduction in cross sectional dimensions of a bar under stretching is described by a number ν, called *Poisson's ratio*. A material loaded along one direction will experience deformation in both that direction and the directions perpendicular to it. This is called the *Poisson effect*, after Siméon-Denis Poisson (1781–1840). If we stretch a bar and get a longitudinal strain ε, then we also get a transverse strain $-\nu\varepsilon$. For many common materials, such as steel, aluminum, or wood, ν is constant and ranges from 0.22 to 0.3.

We consider a homogeneous material where deformation under loading has the same properties in all directions. We have seen that a wire stretches according to Hooke's law. In the present case, we can write this for the x_1-direction as

$$\varepsilon_{11} = \frac{1}{E}\sigma_{11}.$$

But the other stress components affect ε_{11} as well. We consider the linear theory in which the action of any additional load can be simply added (we neglect nonlinear effects). This brings us to the relation

$$\varepsilon_{11} = \frac{1}{E}\left(\sigma_{11} - \nu\sigma_{22} - \nu\sigma_{33}\right).$$

Experiments show that, for practical purposes, the shearing stresses do not affect the normal strains in a homogeneous material; hence the mixed components σ_{12}, σ_{23}, σ_{31} do not affect ε_{11}, and we have a working relation for ε_{11}. Reasoning from symmetry, we can state that for the other nonmixed components we should have

$$\varepsilon_{22} = \frac{1}{E}\left(\sigma_{22} - \nu\sigma_{11} - \nu\sigma_{33}\right),$$

$$\varepsilon_{33} = \frac{1}{E}\left(\sigma_{33} - \nu\sigma_{11} - \nu\sigma_{22}\right).$$

If we wish to find how an elementary volume changes, we add the previous three equations to get

$$\theta = \varepsilon_{11} + \varepsilon_{22} + \varepsilon_{33} = \frac{1 - 2\nu}{E}(\sigma_{11} + \sigma_{22} + \sigma_{33}).$$

Experiment shows that for the shear components the relations are

$$\varepsilon_{ij} = \frac{1}{2G}\sigma_{ij} \qquad (i \neq j),$$

where the *shear modulus* G relates to E and ν through

$$E = 2G(1 + \nu).$$

The shear modulus is analogous to E and acts as a measure of material stiffness under shear loads.

Hooke's equations can be rewritten in another form, this time expressing the stresses in terms of the strains:

$$\sigma_{ij} = \lambda\theta\delta_{ij} + 2\mu\varepsilon_{ij}, \tag{2.29}$$

where δ_{ij} is the Kronecker symbol

$$\delta_{ij} = \begin{cases} 1 & (i = j), \\ 0 & (i \neq j). \end{cases}$$

Here we use *Lamé's notation*[9] for the elastic constants, which are $\mu = G$ and $\lambda = 2G\nu/(1 - 2\nu)$. Note that there are only two independent elastic constants: we can use either the pair E and ν, or the pair λ and μ.

2.13 Constitutive Laws

We have obtained a version of Hooke's law applicable to "homogeneous, isotropic, elastic materials." There are other material models for which the relations between the stress and strain tensors have a more complex form. They can include temperature and other parameters describing the material state. Hooke's law and other such relations are called *constitutive laws* for the material. We discuss here only elastic materials. Let us introduce some terminology.

A material is *homogeneous* if its properties do not depend on the position of the elementary area element used to investigate those properties, and *isotropic* if its properties do not depend on the orientation of the area element. Many materials are *anisotropic*, however. One example is wood, which is composed of fibers that run longitudinally along the "grain." For such materials, Hooke's law relating the tensors of stress and strain takes the more general form

$$\sigma_{ij} = \sum_{k,l=1}^{3} c^{ijkl}\varepsilon_{kl}.$$

---

[9]Gabriel Lamé (1795–1870).

So, each stress component at a point is linearly related to all strain components at the same point. To preserve symmetry of the tensors, the *elastic moduli* c^{ijkl} should satisfy symmetry conditions such as

$$c^{ijkl} = c^{klij} = c^{jikl} = c^{ijlk}$$

for any i, j, k, l taking values from the set $\{1, 2, 3\}$. These requirements leave twenty-one independent constants in the set $\{c^{ijkl}\}$. We should mention that these form the components of a tensor of elastic moduli; because this is an objective quantity for the material, its components must transform according to certain rules when we transform the coordinate frame.

Until now, we have avoided any mention of energy in mechanics. Let us impose the requirement that the following inequality hold for any symmetric tensor ε:

$$\sum_{k,l=1}^{3} c^{ijkl} \varepsilon_{kl} \varepsilon_{ij} \geq c \sum_{i,j=1}^{3} \varepsilon_{ij}{}^2. \tag{2.30}$$

The expression on the left is twice the intensity of the energy due to elastic deformation at a point of the material. It must be positive for any deformation, which (2.30) reflects. We require a bit more: that the constant c is positive and does not depend on ε. The inequality means that the internal energy due to deformation of the body is a positive definite quadratic form, as is required in elasticity. In this way, we define the generalized form of Hooke's law for linear anisotropic elasticity.

But we should understand that this law is not absolute. First of all, there are materials whose behavior under load cannot be described by Hooke's law in principle. In such cases, we need to bring in more complex constitutive laws like the laws of plasticity, viscoelasticity, viscoplasticity, and so on. We will not discuss such relations, but we mention that in them the properties of the material are made to depend on the history of its deformation.

We should also comment on the linearity assumption. For many materials like steel, a linear approximation of the laws and equations is fairly accurate. However, there are materials like copper with inelastic properties for small deformations, and we find that even steel exhibits a nonlinear constitutive equation when we are interested in better than 0.1% accuracy. In such cases, we need to bring in more complex relations. They could contain terms describing temperature effects (because any material will change its temperature a bit during deformation), and should involve corrections to describe process nonlinearities. At some level of applied stress, the linear behavior of a material changes to a largely nonlinear behavior.

2.14 Boundary Value Problems

We have mentioned boundary conditions for various problems. It is now time to discuss how they arise when we treat differential equations.

The straight-line motion of a particle under the action of a force is given by the equation

$$y''(t) = f(t),$$

with $f = f(t)$ given. Its general solution has the form

$$y(t) = c_1 + c_2 t + y_0(t), \tag{2.31}$$

where c_1, c_2 are indefinite constants and $y_0(t)$ is defined by double integration of $f(t)$. To define the c_i uniquely, we normally appoint values of y and y' — the particle's position and velocity — at some time instant. Then we know the whole trajectory. Thus the system

$$y''(t) = f(t), \qquad y(t_0) = y_0, \quad y'(t_0) = v_0,$$

with given y_0, v_0, determines the motion uniquely. Problems of this type are called *initial value*, or *Cauchy, problems*.

However, suppose we take the same equation and change the independent time variable t to a position variable x:

$$y''(x) = f(x).$$

This describes the equilibrium of a wire under a distributed tension load. In terms of x, the wire occupies the interval $[0, L]$. Of course, we could consider a Cauchy problem here as well. This means we could appoint two conditions at the left end, for $y(0)$ and $y'(0)$. At the right-end $x = L$, the force is uniquely determined. However, we know that the value of applied force on the right end is arbitrary and can be easily changed through the application of an additional force. So a Cauchy setup is inappropriate. The reader also understands that neither end of the wire is preferable to the other. Because we have two free constants in (2.31), we must appoint one condition at each end. There are a few ways in which this can be done. The first (yielding the setup for the *first boundary value problem*) is to introduce given displacements at the ends:

$$y''(x) = f(x), \qquad y(0) = a, \quad y(L) = b.$$

This problem always has a unique solution. Because of Hooke's law, when we fix the value of y' at one endpoint we also fix the value of the applied force there. In this way we pose a so-called *mixed boundary value problem*:

$$y''(x) = f(x), \qquad y(0) = a, \quad y'(L) = c.$$

It is clear that we could have specified $y'(0)$ and $y(L)$ instead. The mixed problem always has a unique solution.

We can also specify the forces at both ends:

$$y''(x) = f(x), \qquad y'(0) = c, \quad y'(L) = d.$$

This is the *second boundary value problem*. Observe that no point of the wire is fixed in this case, so the whole thing is free to move. Indeed, it cannot be in equilibrium unless the forces are *self-balanced*. Even with the inclusion of a self-balance condition, a solution will not be unique. A wire free of load, for example, will admit any solution of the form $y(x) = c_1 = $ constant. So any parallel shift in position is admissible, and we would have to fix a point of the wire (e.g., put $y(0) = 0$) in order to specify the solution uniquely. It becomes possible to introduce this third condition (in a problem where the general solution contains only two arbitrary constants!) because of the self-balance requirement.

These boundary value problems for a wire followed from a complete set of relations for the wire. Let us list them in order to see some analogy with the setup of elasticity problems. We have the expression for strain $\varepsilon = \partial y / \partial x$, the stress σ, the constitutive (i.e., Hooke's) law $\sigma = E\varepsilon$, and the equation of equilibrium $S \partial \sigma / \partial x + F = 0$. These reduce to the single equation $y'' = f$ with $f = -F/(ES)$. If we wish to formulate a boundary value problem for the complete system of wire equations with respect to y, ε, σ, we can impose one of the above sets of boundary conditions on y. The situation for elasticity problems is exactly the same: we can formulate them in terms of displacements, for which it is necessary to formulate three boundary conditions at each boundary point. The same boundary conditions should be taken for problems where the setups include displacements and strain and stress tensors.

Let us touch on the issue of boundary conditions for two-dimensional problems. Consider, for example, a square membrane under load. It is described by Poisson's equation

$$u_{xx} + u_{yy} = -F.$$

(Here we use a commonly employed double-subscript notation to denote a second-order partial derivative: $u_{xx} \equiv \partial^2 u / \partial x^2$, for example.) By itself, this cannot determine a unique solution. Indeed, if we seek a purely x-dependent solution $u = v(x)$ (assuming, of course, that F depends only on x as well), then v satisfies

$$v'' = -F.$$

Recognizing this as the equation for the wire, we realize that we should impose a boundary condition at each end $x = 0$ and $x = L$ of the region occupied by the membrane.

The membrane equation has the property of invariance under coordinate rotation: if we rotate the coordinate frame, then the form of the equation remains the same in the new coordinates. This means that no point of the boundary is preferable to any other. But because we found that we must pose one condition at each "endpoint" in the particular case of membrane deformation discussed immediately above, we should expect that, in the general case, exactly one condition must be imposed at each boundary point in order to obtain a well-posed setup. The result of this intuitive reasoning can be verified rigorously. For example, when we appoint values of the displacement u at each point of the membrane boundary, the corresponding boundary value problem, called Dirichlet's problem, has a unique solution. The case when we appoint forces on the membrane boundary is similar to that of the second main problem for the wire.

Now, let us see how these look for a more complex problem of three-dimensional elasticity.

2.15 Setup of Boundary Value Problems of Elasticity

First, let us collect all the relations of linear elasticity. We have three relations for the force components, necessary and sufficient for static equilibrium:

$$\sum_{j=1}^{3} \frac{\partial \sigma_{ij}}{\partial x_j} + F_i = 0 \qquad (i = 1, 2, 3). \tag{2.32}$$

Next, we have equations for small strains:

$$\varepsilon_{ij} = \frac{1}{2} \left(\frac{\partial u_i}{\partial x_j} + \frac{\partial u_j}{\partial x_i} \right) \qquad (i, j = 1, 2, 3). \tag{2.33}$$

Finally, we have Hooke's law:

$$\sigma_{ij} = \lambda \theta \delta_{ij} + 2\mu \varepsilon_{ij} \qquad (i, j = 1, 2, 3). \tag{2.34}$$

With regard for the symmetry present in the stress and strain tensors, this system, defined in the volume V, contains $3 + 6 + 6 = 15$ equations with respect to $6 + 6 + 3 = 15$ unknown functions σ_{ij}, ε_{ij}, u_i. Experience tells us that we have enough equations to find the strain state of a body if we pose some conditions on its boundary ∂V.

We see the parallels to the wire equations of the previous section. Continuing the parallels, we can state that, for a well-defined problem setup, we need to adjoin some boundary conditions. In the previous section, we saw that the sets of boundary conditions were defined for the simplified equation in displacements, and then were used in other circumstances. Let us try to take this route.

First, we write out the equations in displacements, which are derived by elimination of the remaining unknowns. These are Lamé's equations; their derivation can be found in any textbook on elasticity:

$$2\mu\Delta u + \lambda\frac{\partial\theta}{\partial x} + F_1 = 0,$$

$$2\mu\Delta v + \lambda\frac{\partial\theta}{\partial y} + F_2 = 0,$$

$$2\mu\Delta w + \lambda\frac{\partial\theta}{\partial z} + F_3 = 0, \qquad (2.35)$$

where (u, v, w) is the vector of displacement of a body point,

$$\Delta u = \frac{\partial^2 u}{\partial x^2} + \frac{\partial^2 u}{\partial y^2} + \frac{\partial^2 u}{\partial z^2}, \qquad \theta = \frac{\partial u}{\partial x} + \frac{\partial u}{\partial y} + \frac{\partial w}{\partial z}.$$

The Laplacian appears to play an important role in each of these equations. This suggests that we appoint values for each component of the displacement vector at every boundary point:

$$\mathbf{u}\big|_{\partial V} = \mathbf{u}_0. \qquad (2.36)$$

Such a geometrical boundary condition makes sense, and further analysis shows that the boundary value problem for (2.35) and (2.36) is well-posed; that is, it has a unique solution when the external load (F_1, F_2, F_3) is sufficiently smooth along with the boundary values $\mathbf{u}_0$. So we conclude that at each point of the boundary we must appoint one vector boundary condition (equivalently, three scalar conditions). The Lamé equations and the geometrical conditions define the first boundary value problem of linear elasticity.

The wire problem suggests that we could appoint some force conditions on the boundary. In particular, we can appoint three conditions

$$(\sigma_{11}n_1 + \sigma_{12}n_2 + \sigma_{13}n_3)\big|_{\partial V} = f_1,$$

$$(\sigma_{21}n_1 + \sigma_{22}n_2 + \sigma_{23}n_3)\big|_{\partial V} = f_2,$$

$$(\sigma_{31}n_1 + \sigma_{32}n_2 + \sigma_{33}n_3)\big|_{\partial V} = f_3, \qquad (2.37)$$

with given forces (f_1, f_2, f_3). This setup makes mechanical sense, but analysis also shows that the Lamé equations with these boundary conditions define a well-posed problem when all forces acting on the body are self-balanced. However, the solution is defined only up to an infinitesimally small rigid motion of the body as a whole. This also corresponds to our mechanical view of the problem. Here we have the second boundary value problem of elasticity theory.

A variety of other types of boundary value problems can be posed for Lamé's equations. All are said to be mixed. We can, for example, appoint geometrical conditions on part of the boundary, and appoint force conditions on the rest. But there are many other possibilities, and we shall not pursue the issue further.

Now let us return to the complete set of equations. For the wire, we could easily reduce the system to an equation for the stress. The same can be done in elasticity theory: we can eliminate the components of displacement and strain. In this way, we obtain nine equations with respect to σ_{ij}. If we take into account the symmetry of the stress tensor — that is, that these equations contain only six unknown functions $\sigma_{11}, \sigma_{12}, \sigma_{13}, \sigma_{22}, \sigma_{23}, \sigma_{33}$ — then we end up with nine equations in six unknown functions. Nonetheless, analysis shows that this system, when supplemented with three force conditions, presents a well-posed problem having a unique solution when the external forces are self-balanced. We can pose other types of boundary conditions as well, but there must be three at each point of the boundary.

Finally, we can consider how to complete the above fifteen-equation set for the theory of elasticity. In order to pose boundary value problems, we must supplement the equations with three boundary conditions at each point — exactly as was done for Lamé's system.

However, let us discuss these problems a bit. Although they are all equivalent in principle, their structures differ markedly. The Lamé equations turn out to be an elliptic system (in particular, because of (2.30)), and for such systems there are many known results concerning existence of solutions, smoothness (depending on the smoothness of external forces), and so forth. When expressed in terms of stresses, the system is equivalent to the Lamé equations, but looks quite strange: there are nine equations for six unknown functions. This is an example of how dangerous it is to borrow principles directly from linear algebra and assert them in situations where they may not apply. The full setup in terms of fifteen equations inherits all the difficulties of the setup in terms of stresses, but remains equivalent to the Lamé equations, and therefore inherits all the properties of that setup as well.

2.16 Existence and Uniqueness of Solution

We have composed a model of a linearly elastic three-dimensional body. For many common materials, the model works so well that engineers may come to identify it with the real body it is meant to describe. But the model remains a model and cannot be absolutely accurate in principle. The theory of elasticity is a comparatively old science, by which many practical problems have been solved. However, analytical solutions have been found only for a restricted set of problems, while many more have been solved approximately. We should be aware that any approximate solution is a solution to another mathematical model of the same real problem. Strictly speaking, only when error bounds exist for an approximation can we use it to infer properties of the initial model and problem.

Engineers often deem a model satisfactory if its predictions are confirmed by experiment. However, a good mathematical theory must pass more stringent analytical criteria. A complete treatment of a linear problem, in particular, should treat the following crucial issues:

1. *Uniqueness.* Does the problem have just one solution, or more than one?
2. *Existence.* Does the problem have a solution at all?
3. *Continuous dependence on the data.* How does a solution change when we make small changes to various parameters of the problem?

Engineers are normally open to learning about uniqueness and continuous dependence on the data, but they are skeptical about the importance of establishing existence of a solution. This attitude comes partly from the profusion of solutions — both analytical and numerical — to which engineers have been exposed, and partly from their conviction that any real structure should exist in some definite state when placed under load. The latter conviction betrays a fallacious identification between the actual structure and a mathematical model intended to describe it. Numerical solutions also prove little, because they are based on finite dimensional approximations, as we have noted. On more basic logical grounds, the existence of solutions to certain problems simply does not guarantee the existence of solutions to all problems. A good model is characterized by the property that any "sensible" boundary value problem based on the model will have a solution. It took trial and error for people to learn how to properly pose problems in physics, and in the theory of elasticity in particular. We have stated that three boundary conditions must be included in any formulation of a three-dimensional elasticity problem, and that these conditions must be in some sense complementary. At some point on the boundary of a body, for example, we are not free to appoint u_1 and f_1 simultaneously. Why, indeed, should there be three boundary conditions? The answer may be more or less clear for elasticity problems formulated in terms of displacements, but not for problems formulated in terms of stresses. A uniqueness theorem will tell us that no fewer than three boundary conditions should be stipulated, but will not in itself prevent us from imposing more than that.

Even in linear elasticity, there are questions that seem nontrivial. For example, suppose the boundary of a body is under load and the body's motion is unrestricted. It is clear that an arbitrary force can move the body as a rigid whole. What could be said about the solvability of such a problem? We have considered only statics problems, so we could consider a corresponding problem in which the body's mass is zero (i.e., neglect the body's inertia) and thus the set of possible external loads should be restricted. Engineering intuition tells us that the needed restriction involves a condition of *self-balance* of the external forces, giving zero for both the resultant force and moment. In this case, our intuition is correct. But, for the model of an elastic membrane described by Laplace's equation, the self-balance condition reduces to only one equation for the load. Thus, the six self-balance conditions for the membrane as a three-dimensional body reduce to only one for the model.

In other words, existence of a solution is crucial to the workability of a model. Another essential point regarding existence is illustrated by the famous paradox of Perron,[10] which shows us how we can get into trouble by tacitly assuming the existence of a greatest positive integer. Calling this integer N, we have N^2 as a positive integer as well. But $N^2 \leq N$ by assumption, from which it follows that $N = 1$. So an unwarranted supposition of existence can mislead us to the extreme. Despite this, the mathematical challenge associated with existence theorems will cause certain pragmatists to judge them unworthy of attention. In this book, we shall also put the technicalities of existence aside. But the reader should keep in mind that these questions are crucial to the success of a theory. We could further note that existence theorems are available for the problems of linear elasticity, whereas the general question of existence remains open for nonlinear problems.

Another question is whether the solution of a boundary value problem depends continuously on the parameters of the object and the external data. When such continuity (which must be formulated in a rigorous way, in terms of smoothness of the functions involved) holds, the problem is said to be "well-posed." The well-posedness of a problem is normally lost under special circumstances; we refer to this as a *loss of stability*. It occurs with nonlinear models of processes.

Before proceeding to consider uniqueness of solution, we will derive an integral equation that plays important roles on both the theoretical and practical sides of elasticity. This is the *principle of virtual work*, or *principle of virtual displacements*.

Virtual Work Principle

To formulate the principle, we suppose the boundary is composed of two parts: a part S_1, over which condition (2.36) is given, and a part S_2, over which (2.37) is given. Suppose the body is in equilibrium, so that its field of displacements is given by the function $\mathbf{u}(\mathbf{x})$. If we need to change the displacement field by some value $\mathbf{v}(\mathbf{x})$, we must do so in such a way that the geometric condition (2.36) continues to hold. This means that

$$\mathbf{v}\big|_{S_1} = \mathbf{0}. \tag{2.38}$$

Such displacements are called *virtual* (or *admissible*) displacements of the body. It turns out that for any sufficiently smooth admissible displacement field $\mathbf{v}(\mathbf{x})$, the identity

$$\int_V \sum_{i,j=1}^{3} \sigma_{ij}\varepsilon_{ij}(\mathbf{v})\, dV = \int_V \sum_{i=1}^{3} F_i v_i\, dV + \int_{S_2} \sum_{i=1}^{3} f_i v_i\, dS \tag{2.39}$$

holds, where σ_{ij} are the components of the stress tensor for the body in equilibrium and $\varepsilon_{ij}(\mathbf{v}) = \frac{1}{2}(\partial v_i/\partial x_j + \partial v_j/\partial x_i)$ (and, of course, F_i, f_i are the force terms that appeared in equations (2.35) and (2.37)). This identity

---

[10]Oskar Perron (1880–1975).

constitutes the virtual work principle. To establish it, let us multiply the ith equation of (2.32) by v_i, sum over i, and integrate over the domain V:

$$\int_V \sum_{i=1}^{3} \left(\sum_{j=1}^{3} \frac{\partial \sigma_{ij}}{\partial x_j} + F_i \right) v_i \, dV = 0. \tag{2.40}$$

For multiple integrals, we can use the integration-by-parts formula

$$\int_V \frac{\partial g(\mathbf{x})}{\partial x_i} f(\mathbf{x}) \, dV = - \int_V g(\mathbf{x}) \frac{\partial f(\mathbf{x})}{\partial x_i} \, dV + \int_S f(\mathbf{x}) \, g(\mathbf{x}) \, n_i \, dS, \tag{2.41}$$

where f, g are continuously differentiable functions, and n_i is the ith component of the outward unit normal to the boundary S.

Let us integrate by parts in (2.40):

$$\int_V \sum_{i=1}^{3} \left(\sum_{j=1}^{3} \frac{\partial \sigma_{ij}}{\partial x_j} \right) v_i \, dV = - \int_V \sum_{i=1}^{3} \left(\sum_{j=1}^{3} \sigma_{ij} \frac{\partial v_i}{\partial x_j} \right) dV$$
$$+ \int_S \sum_{i,j=1}^{3} \sigma_{ij} v_i n_j \, dS.$$

Next, we can rearrange:

$$\int_V \sum_{i=1}^{3} \left(\sum_{j=1}^{3} \sigma_{ij} \frac{\partial v_i}{\partial x_j} \right) dV = \int_V \sum_{i,j=1}^{3} \sigma_{ij} \varepsilon_{ij}(\mathbf{v}) \, dV.$$

Using (2.38) and (2.37), we get

$$\int_S \sum_{i,j=1}^{3} \sigma_{ij} v_i n_j \, dS = \int_{S_2} \sum_{i=1}^{3} f_i v_i \, dS,$$

so altogether we have proved (2.39).

Why is the name "virtual work principle" applied to this equation? We know from elementary physics that the work done by a force F acting through a distance s is given by the formula $W = Fs$. It is seen that the integrals on the right-hand side of (2.39) represent the work of the external forces F_i and f_i on the corresponding virtual displacements v_i. The term $-\int_S \sum_{i,j=1}^{3} \sigma_{ij} v_i n_j \, dS$ of (2.39) can be explained as the work of internal forces that is represented by the sum of the products of the stresses multiplied by the corresponding strains, because of how the state of the body is specifically described. Quite frequently, then, the virtual work principle is formulated as follows:

The work of the internal and external forces in a body over any virtual displacements is zero.

Before proceeding, we would like to note two important points. First, the derivation of the formula for the virtual work principle did not involve the use of Hooke's law; this means that the principle holds not only for linearly elastic bodies, but for any body that can be described using the tensor of small strains. Second, if we find that the virtual work principle is valid for any smooth virtual displacement, then, as consequences, we have the equations of equilibrium (2.32), and on S_2, the boundary conditions (2.37). A proof of this can be based on the tools of the calculus of variations, and the interested reader can find the proof developed in any textbook on the theory of elasticity. We see that no derivatives of σ_{ij} appear in the virtual work principle, so, formally, this equation requires less smoothness of the stresses than do the initial differential equations. This is the basis for introducing so-called *generalized statements* of the elasticity problem. In this way, modern mathematical elasticity studies the theoretical questions connected with existence, continuity, and the relationship between the smoothness of the external forces and that of the solutions.

Uniqueness

We have seen that uniqueness of solution to a boundary value problem is a key consideration in our acceptance of a mathematical model. Because linear elasticity is a vital model in mechanics, we would like to establish this property for its boundary value problems. The method of proof dates back to Kirchhoff. We remind the reader that if we encounter problems of a linear theory that fail to have unique solutions, then this is evidence of some irregularity in the model; in this case, we need to explain what is happening or, more frequently, to locate bad assumptions in the model.

We would like to demonstrate uniqueness of solution to a general problem of linear elasticity. Our strategy will be to suppose that there are two different solutions $\mathbf{u}_1$ and $\mathbf{u}_2$, and then to demonstrate that their difference $\mathbf{u} = \mathbf{u}_2 - \mathbf{u}_1$ is zero. It is clear that, by linearity of the equations, the difference vector $\mathbf{u}$ must satisfy any equation satisfied by the solutions, but with zero external forces and zero boundary conditions. Thus, on S_1, we get $\mathbf{u} = \mathbf{0}$, and the equation of the virtual work principle takes the form

$$\int_V \sum_{i,j=1}^3 \sigma_{ij}(\mathbf{u})\varepsilon_{ij}(\mathbf{v})\, dV = 0.$$

Here, the notation $\sigma_{ij}(\mathbf{u})$ indicates that σ_{ij} must be calculated through the difference vector $\mathbf{u}$.

This equality holds for any admissible $\mathbf{v}$; because $\mathbf{u}$ is admissible, we can set $\mathbf{v} = \mathbf{u}$ and get

$$\int_V \sum_{i,j=1}^3 \sigma_{ij}(\mathbf{u})\varepsilon_{ij}(\mathbf{u})\, dV = 0.$$

Now, we bring in Hooke's law in the generalized form

$$\sigma_{ij} = \sum_{k,l=1}^3 c^{ijkl}\varepsilon_{kl},$$

getting

$$\int_V \sum_{i,j,k,l=1}^3 c^{ijkl}\varepsilon_{kl}\varepsilon_{ij}\, dV = 0, \qquad \varepsilon_{ij} = \varepsilon_{ij}(\mathbf{u}).$$

Inequality (2.30) shows that

$$\int_V \sum_{i,j=1}^3 \varepsilon_{ij}{}^2\, dV = 0,$$

and thus over the domain V, we have $\varepsilon_{ij} = 0$ or, equivalently,

$$\varepsilon_{ij} = \frac{1}{2}\left(\frac{\partial u_i}{\partial x_j} + \frac{\partial u_j}{\partial x_i}\right) = 0 \qquad \text{for all } i, j = 1, 2, 3.$$

This system of equations has no solutions other than those of the form

$$\mathbf{u} = \mathbf{a} + \mathbf{b} \times \mathbf{x},$$

which represent vectors describing infinitely small rigid motions of the corresponding rigid body. Because of the condition on S_1, which is $\mathbf{u} = \mathbf{0}$, we conclude that $\mathbf{u} = \mathbf{0}$ throughout V.

This proof, given by Kirchhoff before the end of the nineteenth century, was a real achievement. Prior to that time, engineers solved particular problems and had to produce a practical argument in favor of uniqueness of solution in each specific case.

As a final point, we would like to emphasize that any model of a real process is derived under certain conditions. It therefore inherits some properties of the real process, while failing to account for others. The properties of a real process we see in nature should stimulate our interest in verifying whether these have found accurate expression in the model. Among the central points we must consider when judging the validity of a model are the properties of uniqueness of solution, existence of solution (with accompanying information regarding restrictions on external and internal system parameters), and qualitative behavior of the solution.

2.17 Energy; Minimal Principle for a Spring

In the previous section, we dealt with the concept of work. This is a fundamental notion of mechanics and of physics as a whole. The product Fs gives a single number W that serves to characterize the action of a force during some process involving movement. Work is a quantity that can only be calculated, but it is nonetheless important and we need to consider it more carefully. If the force changes during its action so that F depends on the length parameter x, then we cannot use the above formula. Instead, we must subdivide the distance into infinitely small parts (as usual, this is done with differentials dx), calculate the work $F(x)\,dx$ of the force over each infinitely small interval, and sum up these elementary portions of the work. This gives us an integral representation for the work of the force:

$$W = \int_{x_0}^{x_1} F(x)\,dx. \tag{2.42}$$

When the directions of a force $\mathbf{F}$ and distance $\mathbf{s}$ differ, then the work of $\mathbf{F}$ is defined by its projection onto $\mathbf{s}$: $W = \mathbf{F} \cdot \mathbf{s}$. When $\mathbf{F}$ acts along a curve L specified by $\mathbf{x} = \mathbf{x}(s)$, where s is the length parameter, the work can be written as

$$W = \int_L \mathbf{F}(s) \cdot d\mathbf{x} = \int_L \mathbf{F}(s) \cdot \mathbf{x}'(s)\,ds.$$

Related to work is another central notion of physics: energy. This quantity also serves to characterize a process in total. Students in elementary physics often have rather fuzzy notions about energy, thinking of it as something that magically changes its form while allowing itself to be neither created nor destroyed. So they watch its appearance in the form of work: a furnace heating a house, a television set emitting light and sound, a car moving, and so on. The notion of energy was elaborated for several centuries, and the first law of energy conservation was eventually established for elementary problems in which only the motion of bodies was involved. For any such simple system, it was found that we can equate the sum of the kinetic and potential energies at some time instant to the work of all external forces that have acted on the system. Later, it was discovered that the work of forces can be transformed into the heating of liquids such as water, and that the rise in temperature of a given volume of liquid can be calculated precisely. We now know that it is impossible to design an engine that provides more energy than it needs to operate. The history of physics is littered with attempts to find exceptions to this, but the nonexistence of a "perpetual motion machine" remains an absolute postulate.

We cannot discern the total energy stored in a system; much energy is hidden, such as the enormous energy we see released during the explosion of an atomic bomb. However, it often suffices to consider only large-scale phenomena such as motion, deformation, and heating. This fact underlies how we generally treat many problems in both elasticity and continuum mechanics.

Let us consider the energy problem for a simple spring. Here, the relation between a stretching force F and the resulting displacement x of the spring's endpoint is given by the formula

$$F = kx, \tag{2.43}$$

where k is the spring *stiffness constant*. This is simply a version of Hooke's law. First, we wish to learn how much work must be expended in order for a force to elongate the spring by some length x. We might be tempted to apply a constant force F and multiply it by x, but this leads to immediate difficulties. The process of stretching an initially relaxed spring involves the inertia of the spring: the spring presents no resistance to elongation at first (when $x = 0$), and therefore the work of the force F goes into the kinetic energy associated with acceleration of the spring. As we wish to avoid such considerations, we choose to maintain a static picture of the entire deformation; we do this by starting F at an initial value of zero and increasing it infinitely slowly.[11] We therefore apply (2.42) and let the force increase gradually according to (2.43). When the displacement of the spring end is ξ and the corresponding force is $f = k\xi$, then the additional work needed to produce an infinitesimal displacement $d\xi$ is, up to quantities of the second order of smallness, $dW = k\xi \, d\xi$; hence, the total work needed to lengthen the spring statically by an amount x from the nondeformed state is given by

$$W = \int_0^x k\xi \, d\xi.$$

After integration, we get

$$W = \frac{kx^2}{2} = \frac{Fx}{2}.$$

So, we got exactly half the answer that we first expected. This quantity is called the *internal energy* of the spring. If we lengthen the spring by an amount x and then allow it to relax very slowly, we can get the work $Fx/2$ back again. We can therefore regard $kx^2/2$ as energy stored during

---

[11]Such *quasistatic assumptions* about motion are common in physical reasoning. In reality, a body cannot move between two different stationary positions without some acceleration coming into the picture. The "infinitely slow" assumption is based on the fact that we can neglect the inertial forces inherent in a process that occurs very slowly. Some caution is required with this kind of reasoning, however, because certain integral characteristics of a process can remain the same no matter how slowly the process is made to occur (an example of this is the distance x in the present case). Infinite slowness in physics is really practical slowness: in many mechanics experiments, for instance, we could achieve the desired effect by spreading the process out over a couple of minutes.

the deformation, and refer to it as *potential energy*. Let us denote it by E:

$$E = \frac{kx^2}{2}. \tag{2.44}$$

Note that if we started with a nonlinear relation between F and x for the spring, then the result would be different. In particular, the factor of $1/2$ is common to all *linear* theories for potential energy. It is easy to see that

$$\frac{\partial E}{\partial x} = kx = F. \tag{2.45}$$

We therefore refer to E as a *potential function* for the force F.

Now, let P be the work of the force F_0 over the displacement x that corresponds to the force F_0. Thus, $F_0 = kx$, $P = F_0 x$. This P is, besides, the potential of the force F_0. Another simple fact is that the expression

$$\Phi = E - P = \frac{kx^2}{2} - F_0 x$$

takes its minimum value when $x = F_0/k$, so the minimum is taken at the equilibrium position of the spring. This is called the *principle of energy minimum*. In the next section, we will see corresponding statements for a three-dimensional elastic body.

As a final note, we could take the expression $kx^2/2$ for E and, using the explanation above, "derive" Hooke's law for the spring. Because the role of strain is played by x here, we could present a theory of the spring, beginning with its expression for internal potential energy. The same is possible in general elasticity; many textbooks begin their presentations of elasticity by formulating the potential energy density for a three-dimensional body. From this, they derive the generalized Hooke's law and the other main statements of the theory. We gave preference to its historical development; this was longer, but historically oriented developments usually make clearer the actual thought processes that were used to develop the principal notions of a subject.

2.18 Energy in Linear Elasticity

We now extend the ideas of the previous section to the problems of three-dimensional linear elasticity. We begin with the work of forces over part of an elastic body. Let us repeat some arguments we have used deriving the virtual work principle. By analyzing the derivation, we see that the same steps applied to an arbitrary volume $\mathcal{V}$ (with boundary S) of the elastic body give us the equality

$$\int_{\mathcal{V}} \sum_{i,j=1}^{3} \sigma_{ij}\varepsilon_{ij}(\mathbf{v})\, d\mathcal{V} = \int_{\mathcal{V}} \sum_{i=1}^{3} F_i v_i \, d\mathcal{V} + \int_{S} \sum_{i,j=1}^{3} \sigma_{ij} n_j v_i \, dS$$

for any admissible displacement $\mathbf{v}$. On the right-hand side, we see the work of external forces: the external volume forces are the F_i, and the external surface forces $\sigma_{ij}n_j$ represent the reactions of that portion of the body external to $\mathcal{V}$. On the left-hand side, we see a volume integral, and our arguments of the previous section lead us to identify this integral as the incremental potential energy of deformation associated with the infinitely small displacement field $\mathbf{v}(\mathbf{x})$ (the smallness assumption is necessary in order for us to neglect additional terms that would be due to the change of stresses inside the body).

Now, let us use this equality to get the full energy stored in $\mathcal{V}$ during deformation of the body under external load. By the same arguments as for the spring, we need to calculate the work of external forces when they change from zero to the final values $F_i(\mathbf{x})$, $f_i(\mathbf{x})$. The uniqueness theorem tells us that the final state of the body does not depend on how we change these values. This does not mean that the energy must be the same for any law of change; however, this could be shown to be the case afterwards. This nondependence of potential energy on the way in which forces are increased to their final values forms the basis for one of the many possible definitions of the term "elastic body." We take the law of change for the forces (and the given values of the displacement field on the boundary) to be one of proportionality to a parameter t, which, in turn, varies from 0 to 1. If $\mathbf{u}(\mathbf{x})$ is the final displacement field, then for any t, the displacement is $t\mathbf{u}(\mathbf{x})$ and, similarly, by linearity of the problem, the components of the stress tensor are $t\sigma_{ij}$. Now, in the state $t\mathbf{u}$, we can consider as infinitely small an admissible disturbance $dt\,\mathbf{u}$, so we have

$$\int_{\mathcal{V}} \sum_{i,j=1}^{3} t\sigma_{ij}\,\varepsilon_{ij}(\mathbf{u}\,dt)\, d\mathcal{V} = \int_{\mathcal{V}} \sum_{i=1}^{3} tF_i\, u_i\, dt\, d\mathcal{V} + \int_{S} \sum_{i,j=1}^{3} t\sigma_{ij} n_j\, u_i dt\, dS.$$

Because we wish to find the whole value, we need to integrate over t from 0 to 1. The integration gives us the total work of forces, and thus the stored internal energy (on the left):

$$E(\mathcal{V}) = \frac{1}{2} \int_{\mathcal{V}} \sum_{i,j=1}^{3} \sigma_{ij}\varepsilon_{ij}\, d\mathcal{V}$$

$$= \frac{1}{2} \int_{\mathcal{V}} \sum_{i=1}^{3} F_i u_i\, d\mathcal{V} + \frac{1}{2} \int_{S} \sum_{i,j=1}^{3} \sigma_{ij} n_j u_i\, dS. \qquad (2.46)$$

We see that inside the volume V there is defined a density of the internal energy, which is

$$e = \frac{1}{2} \sum_{i,j=1}^{3} \sigma_{ij} \varepsilon_{ij}. \tag{2.47}$$

Introducing Hooke's law into this, we have

$$e = \frac{1}{2} \sum_{i,j,k,l=1}^{3} c^{ijkl} \varepsilon_{kl} \varepsilon_{ij}.$$

Partial differentiation (taking into account the symmetry of the elastic moduli tensor) gives us

$$\sigma_{ij} = \frac{\partial e}{\partial \varepsilon_{ij}},$$

which is analogous to (2.45). We could begin elasticity theory from this point and simply postulate the form of potential energy, which we could then use to derive Hooke's law. Our next subtopic confirms that this approach allows us to derive the equilibrium equations as well.

Principle of Minimum Energy

We see that the internal energy due to deformation of a three-dimensional body is given by the expression

$$\mathcal{E}(\mathbf{u}) = \frac{1}{2} \int_V \sum_{i,j,k,l=1}^{3} c^{ijkl} \varepsilon_{kl} \varepsilon_{ij} \, dV.$$

The virtual work principle states that the elastic body is described by (2.39):

$$\int_V \sum_{i,j,k,l=1}^{3} c^{ijkl} \varepsilon_{kl}(\mathbf{u}) \varepsilon_{ij}(\mathbf{v}) \, dV = \int_V \sum_{i=1}^{3} F_i v_i \, dV + \int_{S_2} \sum_{i=1}^{3} f_i v_i \, dS, \tag{2.48}$$

which holds for any admissible displacement field $\mathbf{v}$ that is sufficiently smooth and vanishes on S_1. If (2.48) holds for any admissible displacement $\mathbf{v}$, then $\mathbf{u}$ is a unique solution to the boundary value problem.

We would like to relate this to the principle of minimum energy for the body. For a spring, we found that the minimum of the function $E - F_0 x = kx^2/2 - F_0 x$ gave us the equilibrium equation. In mechanics, the term "functional" is often used for expressions like $\mathcal{E}(\mathbf{u})$ that depend integrally on the values of certain parameters. This dependence extends the notion of function, because a functional maps functions representing the fields inside some domain into real numbers. So let us consider the functional

$$\Phi(\mathbf{u}) = \mathcal{E}(\mathbf{u}) - \int_V \sum_{i=1}^{3} F_i u_i \, dV - \int_{S_2} \sum_{i=1}^{3} f_i u_i \, dS.$$

This energy functional takes the value $-\frac{1}{2}\mathcal{E}(\mathbf{u})$ at an equilibrium state $\mathbf{u}$. The functional $\mathcal{E}(\mathbf{u})$ is nonnegative and second-order homogeneous in the u_i, whereas the terms for the work of external forces F_i, f_i are linear in u_i; this suggests that the behavior of $\Phi(\mathbf{u})$ is similar to that of a nonnegative quadratic function. Because of this, we expect it to grow with increasing $\mathbf{u}$. This means that its minimum value is taken in a bounded ball $|\mathbf{u}| < R$. Thus, there should exist a displacement field $\mathbf{u}$ at which $\Phi(\mathbf{u})$ takes its minimal value. Let us suppose the existence of such a displacement field $\mathbf{u}$. Then

$$\Phi(\mathbf{u} + t\mathbf{v}) \geq \Phi(\mathbf{u})$$

for any admissible $\mathbf{v}$ and any parameter t. But

$$\Phi(\mathbf{u} + t\mathbf{v}) = \mathcal{E}(\mathbf{u}) + t \int_V \sum_{i,j,k,l=1}^{3} c^{ijkl} \varepsilon_{kl}(\mathbf{u})\varepsilon_{ij}(\mathbf{v}) \, dV + t^2 \mathcal{E}(\mathbf{v})$$

$$- \int_V \sum_{i=1}^{3} F_i u_i \, dV - \int_{S_2} \sum_{i=1}^{3} f_i u_i \, dS$$

$$- t \left(\int_V \sum_{i=1}^{3} F_i v_i \, dV - \int_{S_2} \sum_{i=1}^{3} f_i v_i \, dS \right),$$

and thus, by the previous inequality, we get

$$t \left[\int_V \sum_{i,j,k,l=1}^{3} c^{ijkl} \varepsilon_{kl}(\mathbf{u})\varepsilon_{ij}(\mathbf{v}) \, dV \right.$$

$$\left. - \int_V \sum_{i=1}^{3} F_i v_i \, dV - \int_{S_2} \sum_{i=1}^{3} f_i v_i \, dS \right] + t^2 \mathcal{E}(\mathbf{v}) \geq 0.$$

This inequality has the form $at + bt^2 \geq 0$ and can hold for any t if and only if $a = 0$:

$$\int_V \sum_{i,j,k,l=1}^{3} c^{ijkl} \varepsilon_{kl}(\mathbf{u})\varepsilon_{ij}(\mathbf{v}) \, dV - \int_V \sum_{i=1}^{3} F_i v_i \, dV - \int_{S_2} \sum_{i=1}^{3} f_i v_i \, dS = 0.$$

This coincides with (2.48), and thus the minimum value of $\Phi(\mathbf{u})$ can be taken only on the solution of the boundary value problem determined by the virtual work principle. Thus, the problem of finding the minima of the energy $\Phi(\mathbf{u})$ gives us a unique solution to the boundary value problem. The minimum formulation of the problem of equilibrium for an elastic body is equivalent to its statement with the virtual work principle. The use of the minimum energy principle allows us to formulate various approximate methods for solving problems of elasticity, such as the finite element method that is widely used in engineering calculations.

Despite our statement on the equivalence of the virtual work and minimum energy principles in linear elasticity, they are not equivalent in general continuum mechanics. There are nonelastic mechanical models of bodies for which the virtual work principle functions properly, while the minimum energy principle cannot be formulated because an expression for internal energy is lacking. Moreover, in the framework of the virtual work principle, we can consider external loads that depend on the displacements of the body, and may not be potential loads that can be defined with the use of potential functions.

2.19 Dynamic Problems of Elasticity

So far, we have considered statics problems. Dynamics is no less important, however, and we are fortunate that the equilibrium equations can be used to formulate equations for nonequilibrium situations. In classical mechanics, there is a principle due to Jean d'Alembert (1717–1783), who proposed a way in which a dynamics problem can be considered as a kind of statics problem. That is, having Newton's second law $F = ma$ for a material point, we can introduce a so-called *force of inertia* given by $F_1 = -ma$. Then the equation of dynamics appears exactly like the equation of equilibrium for the same material point when the latter is subjected to two forces, F and F_1, having a common line of action. This was generalized to systems of material points and to solid bodies: instead of terms involving acceleration, we introduce formal inertia forces that, together with the external forces, give us the equilibrium equations for a mechanical system. This can be extended to elasticity problems: besides the volume forces $\mathbf{F}_i$ acting on an elastic body, we introduce the inertia forces $-\rho \partial^2 \mathbf{u}/\partial t^2$, where ρ is the density of the material. The sum total of all forces present must keep the body in equilibrium. Thus, in component form, the Lamé equations for statics (2.35) lead us to the dynamic equations

$$2\mu\Delta u_i + \lambda\frac{\partial\theta}{\partial x_i} = \rho\frac{\partial^2 u_i}{\partial t^2} - F_i \qquad (i = 1, 2, 3). \qquad (2.49)$$

To properly pose a problem for these equations, we need to formulate boundary conditions. Specifically, we need conditions that coincide with the combined conditions (2.36) and (2.37) of the static problem, and we must also stipulate conditions on the initial location and velocity field of the body:

$$\mathbf{u}\big|_{t=0} = \mathbf{u}_0, \qquad \frac{\partial \mathbf{u}}{\partial t}\bigg|_{t=0} = \mathbf{v}_0. \tag{2.50}$$

These are appointed on the grounds that if we consider states that do not depend on spatial coordinates, we obtain equations looking like Newton's ordinary equations of motion, for which initial conditions of the type (2.50) are natural. Of course, this only supports the possibility for such a problem setup. The real confirmation is the proof that such problems really have a unique solution. All together, this allows us to pose dynamics problems that have unique solutions belonging to certain classes of functions. (It is clear that it would be possible to formulate conditions different from (2.50) — for example, to specify values of $\mathbf{u}$ at two time instants. But such problems are ill-posed and do not appear in practice.) Few of these problems have analytic solutions. Series solutions are also restricted to a few types of problems, so we are left with numerical methods as the main method of attack. Unfortunately, four independent variables x_1, x_2, x_3, t are involved. Thus, if we introduce a grid with 100 nodes along each coordinate, then we get 10^8 grid nodes, and hence this many nodes for each of our unknown functions u_i. Such massive calculations are rarely undertaken in practice. Instead of solving three-dimensional dynamics problems, therefore, we often solve problems involving steady-state oscillations of a body so that time is eliminated as an independent variable. In such cases, we seek solutions of the form

$$\mathbf{u}(\mathbf{x}, t) = \mathbf{v}(\mathbf{x})\sin(\omega t + \varphi).$$

We also suppose that the external forces are proportional to the same sinusoid: $F_i(\mathbf{x}, t) = \Phi_i(\mathbf{x})\sin(\omega t + \varphi)$. Substituting this into (2.49) and canceling the factor $\sin(\omega t + \varphi)$, we get

$$2\mu\Delta v_i + \lambda\frac{\partial\theta(\mathbf{v})}{\partial x_i} = -\rho\omega v_i - \Phi_i \qquad (i = 1, 2, 3). \tag{2.51}$$

Together with the boundary conditions, this gives us a boundary value problem for the amplitude of the oscillations $\mathbf{v}$.

The *free oscillations* of the same body are described by

$$2\mu\Delta v_i + \lambda\frac{\partial\theta(\mathbf{v})}{\partial x_i} = -\rho\omega v_i \qquad (i = 1, 2, 3).$$

It is clear that this problem has zero as a solution. However, there are values of ω for which nontrivial solutions appear: these are the *resonant frequencies* of the body. At any such frequency, also known as an *eigenfrequency* or *eigenvalue* of the problem, the system comes into *resonance* (meaning that the amplitude of its oscillations will increase without bound if the force is periodic with this period). In the present case, there are also periodic forces at the resonant frequency that fail to excite resonances of the body. The reader should also understand that no real oscillation can grow to infinity: the prediction of such is a consequence of the linear model. In fact, for some amplitude of oscillation, we need to introduce nonlinear effects that restrict the oscillation from growing further. In many cases, forced oscillation at an eigenfrequency will result in a damaged structure. We shall discuss resonance in what follows.

2.20 Oscillations of a String

Let us revisit the elasticity equations we have derived. In all stages of the derivation we have employed limit passages. We did this when we established the properties of stress and the equations of equilibrium. In addition, we used the notion of small deformation. We acted as though the deformation was infinitely small, and thereby neglected terms of the second order of smallness in the derivatives of the displacement; however, the equations were then applied to cases in which the displacements and their derivatives were small but finite. This situation is common in mathematical physics, where terms are often neglected if they are small in comparison with other terms in the same equation. Linear elasticity is a typical linear model of mechanics.

We would now like to discuss another linear model: that describing the transverse oscillations of a string. It is somewhat curious that this model, being linear, describes a nonlinear phenomenon. Let us suppose that the string is stretched tightly with a force T and, in addition, that a distributed load $f(x, t)$ is applied transversely. We considered Hooke's law for a wire under tension; in that case the law was linear, but it becomes nonlinear for deformations that are not small. For a string made from twisted fibers, it is evident that the constitutive relation is nonlinear. Indeed, in the early stages of stretching, a relatively small force can cause significant elongation; in the later stages, the same force cannot. However, we can describe the transverse motions of the string if we know the force T. In this way, we can find the eigenfrequencies of the string knowing only T, the length L, and the mass density ρ. Therefore, the eigenfrequencies of a string do not depend on the material out of which it is made. This may seem to imply that we could make a good violin out of rubber bands. We will be incorporating other assumptions into our model: for example, that the string is *perfectly flexible* and offers no bending resistance. We similarly neglect the string thickness and air resistance. More importantly, though, the sound of a violin is intimately related to the structure of its wooden body.

Henceforth we shall use the term "string" for an object whose transverse displacement is given by a function $u(x, t)$. As in linear elasticity, we neglect terms involving $u_x{}^2$ (by the indices x and t, we denote partial derivatives with respect to these variables), which are small in comparison with unity. We suppose that u has continuous partial derivatives up to second order.

First, we show that the tension in the string remains the same under this assumption. Indeed, the length of any piece $[x_1, x_2]$ of the string becomes, after deformation,

$$\int_{x_1}^{x_2} \sqrt{1 + u_x{}^2}\, dx \approx x_2 - x_1.$$

This means that the stretching force, which depends only on the change in length, remains the same after deformation — at least to within the accuracy of the model. Before deformation, the string is straight and its stretching force is therefore constant; this latter condition persists after deformation as well.

So we see that, in this model, the elastic properties of the string are provided not by elasticity of the material, but simply by stretching: a deflection of the string introduces a vertical projection of the stretching force, and this *restoring force* tends to return the string to its nondeflected shape. Let us find this projection. If α is the angle of the tangent to $u = u(x, t)$ at (x, t), then the vertical projection of the stretching force is $T \sin \alpha$. We know that $\tan \alpha = u_x$. Then, within our chosen accuracy, the projection is

$$T \sin \alpha = T u_x.$$

In order to derive the equation of oscillation, we will apply Newton's second law in the form

$$mv(t_2) - mv(t_1) = \int_{t_1}^{t_2} F\, dt.$$

This is the integral form for linear momentum. We consider a portion of the string $[x_1, x_2]$ over a time interval $[t_1, t_2]$. The vertical component of linear momentum for this portion is

$$\int_{x_1}^{x_2} \rho(x) u_t(x, t)\, dx,$$

and thus the increment of the linear momentum over the time interval of interest is

$$\int_{x_1}^{x_2} \left(\rho(x) u_t(x, t_2) - \rho(x) u_t(x, t_1) \right) dx = \int_{x_1}^{x_2} \rho(x) \int_{t_1}^{t_2} u_{tt}(x, t)\, dt\, dx.$$

The impulse of all the "external" forces applied to this part of the string consists of the impulse of the transverse force

$$\int_{t_1}^{t_2} \int_{x_1}^{x_2} f(x, t)\, dx\, dt$$

and the impulse of the vertical reaction of the rest of the string,

$$\int_{t_1}^{t_2} T\left(u_x(x_2, t) - u_x(x_1, t)\right) dt = \int_{t_1}^{t_2} T \int_{x_1}^{x_2} u_{xx}(x, t) \, dx \, dt.$$

Thus, Newton's second law in momentum form becomes

$$\int_{x_1}^{x_2} \int_{t_1}^{t_2} \rho(x) u_{tt}(x, t) \, dt \, dx = \int_{t_1}^{t_2} \int_{x_1}^{x_2} T u_{xx}(x, t) \, dx \, dt$$
$$+ \int_{t_1}^{t_2} \int_{x_1}^{x_2} f(x, t) \, dx \, dt.$$

To obtain a differential equation from this, we shall use the mean value theorem for integrals, which states that the equality

$$\int_a^b g(x) \, dx = g(\xi)(b - a)$$

holds for some $\xi \in [a, b]$ if $g(x)$ is continuous on $[a, b]$. Applying this to each of the integrals, we have

$$\rho(x) u_{tt}(x, t) \Big|_{\substack{x=\xi_1 \\ t=\tau_1}} = T u_{xx}(x, t) \Big|_{\substack{x=\xi_2 \\ t=\tau_2}} + f(x, t) \Big|_{\substack{x=\xi_3 \\ t=\tau_3}}$$

for some intermediate values ξ_i and τ_i. If we take $x_1 \to x$, $x_2 \to x$, $t_1 \to t$, and $t_2 \to t$, then the equation becomes

$$\rho u_{tt} = T u_{xx} + f.$$

Suppose ρ does not depend on x. Introducing $a = T/\rho$ and $F = f/\rho$, we get the final result:

$$u_{tt} = a u_{xx} + F. \tag{2.52}$$

This is the equation of oscillation of the string. It is the simplest hyperbolic differential equation. We need to supplement it with some boundary conditions: for example,

$$u(0, t) = u_0(t), \qquad u(L, t) = u_1(t).$$

The initial-boundary value problem requires, in addition, the conditions

$$u(x, 0) = v_0(x), \qquad u_t(x, 0) = v_1(x), \qquad x \in [0, L].$$

To get the eigenfrequencies of the string, we consider those solutions of (2.52) that have the form $u(x, t) = v(x) \sin(\omega t + \phi)$. Putting $F = 0$, we arrive at the equation

$$v_{xx} = -\omega^2 v,$$

whose nonzero solutions (for homogeneous conditions $v(0) = 0 = v(L)$) give us the spectrum of the string. The reader is encouraged to work through the details.

Let us emphasize that the equation we have obtained does not suppose a linear Hooke's law for the string. In fact, we have considered one of a wide class of problems with a prescribed initial stress state; the way in which we have derived the problem statement could be referred to as the *linearization* of a nonlinear problem. Linearization plays an important role in mechanics and mathematical physics in general. A linearized problem will often inherit some information about the initial nonlinear problem through the coefficients that remain. In the present problem, this statement applies to the relation between the spring tension T and the density ρ. Problems obtained by linearization can describe many everyday objects, although the case of a three-dimensional body presents additional difficulties. The nicest example of a prestressed body is a crystal vase. When molten glass cools, it solidifies, but temperature differentials present during cooling can build stresses into the solid body. These are not small, and their occurrence is hard to prevent. This is why many vases can shatter with a single touch (or sometimes less stimulus than that). Various methods have been developed to circumvent the formation of built-in stresses in bodies. Sometimes, however, such stresses are introduced intentionally, as is the case with prestressed concrete. We shall discuss this later.

Equations describing the elastic membrane (Laplace's equation for statics or the wave equation for dynamics) can be based on the same sorts of hypotheses we used for the string. A smallness assumption on transverse deflections involves the fact that the tension in the membrane remains the same during deformation. We shall not discuss this further, but the reader can find a development in any textbook on mathematical physics.

2.21 Lagrangian and Eulerian Descriptions of Continuum Media

In describing the statics of an elastic body, we have referred all of its points to its initial state. This was possible because we supposed the strains to be small. Nonsmall displacements are possible under this assumption, however, and are associated with rigid rotations of parts of the body that need not be small when the strains are small. An example is the bending of a thin strip (we used a ruler), for which the strains remain small but the deflection can be significant. When we consider deformational dynamics, we suppose that both strains and displacements are small. In this case, we also refer everything to the initial state of the body. We could refer the points of a body to any state — not necessarily the initial state — and write down equations whose solutions track the location of any chosen material point within the body. This approach is said to be *Lagrangian*.[12] Of course, it is not a good way to describe what happens over time at some particular location in a river: if we fix a material point, it leaves the location of interest sooner or later. Besides, the deformation for river flow would be very large if we were to attempt to describe the river as we would an elastic body. So we need another approach. In the *Eulerian approach* (actually developed before Euler's time) we use spatial coordinates unrelated to certain points of the body: when we fix $\mathbf{x}$, we fix a location in space that can be occupied by different points of the body at different times. So, unlike the Lagrangian approach in which we use $\mathbf{x}$ to refer to a particular point of the body that can move through space, in the Eulerian approach, we use $\mathbf{x}$ to refer to a particular location in space at which some point of the body will be present at some particular time instant (subsequently the material point can move to another location not described by the same value of $\mathbf{x}$). The Eulerian description is appropriate for fluid flow. The model of a fluid, derived in the next section, will be capable of preserving everything we know about the flow properties of liquids and gases. For this model, we will take as unknown variables the field of velocities instead of studying the displacements of material points. Thus, we introduce a field $\mathbf{v} = (v_1, v_2, v_3)$ that specifies the flow velocity at any given spatial location (x_1, x_2, x_3) and time t.

Accelerations should appear in the equations of continuum dynamics. Let us note carefully that in the Eulerian approach we cannot calculate the acceleration of a material point by simply taking the partial derivative of the velocity field with respect to t: we would get the rate of change of velocity at a fixed spatial location, but such a location could be occupied by two distinct material particles at two (even very close) time instants. To find the increment of velocity for a material point over a time interval Δt, we need to consider *that same material point* at times t and $t + \Delta t$; that is, we need to "follow the motion" when calculating this increment. Because we will need to do this for other quantities as well, let us consider an abstract function $G(\mathbf{x}, t)$. Again, we seek the rate of change of G with respect to t as seen by a material point that moves past the location $\mathbf{x}$ at time t. We need what is known as a "total derivative" of G with respect to t: this takes into account both the dependence of G on t *and* the changing conditions experienced by a material point by virtue of its motion through space. At time $t + \Delta t$, our material point has moved to location $\mathbf{x} + \mathbf{v}(\mathbf{x})\Delta t$ (up to terms of the second order of smallness with respect to Δt) and thus experiences a value of G approximately equal to $G(\mathbf{x} + \mathbf{v}(\mathbf{x})\Delta t, t + \Delta t)$ at time $t + \Delta t$. The increment

---

[12] Joseph-Louis Lagrange (1736–1813). However, the approach was applied long before he used it.

of G following the motion is therefore

$$\Delta G = G(x_1 + v_1\Delta t, x_2 + v_2\Delta t, x_3 + v_3\Delta t, t + \Delta t) - G(x_1, x_2, x_3, t)$$
$$= \frac{\partial G(x_1, x_2, x_3, t)}{\partial x_1}v_1\Delta t$$
$$+ \frac{\partial G(x_1, x_2, x_3, t)}{\partial x_2}v_2\Delta t$$
$$+ \frac{\partial G(x_1, x_2, x_3, t)}{\partial x_3}v_3\Delta t$$
$$+ \frac{\partial G(x_1, x_2, x_3, t)}{\partial t}\Delta t + \left(\begin{array}{c}\text{terms of higher order} \\ \text{of smallness in } \Delta t\end{array}\right).$$

Dividing this by Δt and producing the limit passage as $\Delta t \to 0$, we get

$$\frac{DG}{Dt} = \lim_{\Delta t \to 0}\frac{\Delta G}{\Delta t} = \frac{\partial G}{\partial t} + \frac{\partial G}{\partial x_1}v_1 + \frac{\partial G}{\partial x_2}v_2 + \frac{\partial G}{\partial x_3}v_3. \qquad (2.53)$$

We call this the *total*, or *complete, derivative* of G with respect to t. Note that if we introduce simultaneously the Lagrangian space coordinates that refer to points of the body, then this total derivative DG/Dt is the partial derivative in time t of the function G of the Lagrangian coordinates at the point $\mathbf{x}$ that coincides with the given spatial point at time t.

Let us apply (2.53) to each velocity component to get the material acceleration at location $\mathbf{x}$ and time t:

$$a_i = \frac{Dv_i}{Dt} = \frac{\partial v_i}{\partial t} + \frac{\partial v_i}{\partial x_1}v_1 + \frac{\partial v_2}{\partial x_2}v_2 + \frac{\partial v_3}{\partial x_3}v_3. \qquad (2.54)$$

Hence the acceleration vector can be expressed as

$$\mathbf{a} = \frac{\partial \mathbf{v}}{\partial t} + \frac{\partial \mathbf{v}}{\partial x_1}v_1 + \frac{\partial \mathbf{v}}{\partial x_2}v_2 + \frac{\partial \mathbf{v}}{\partial x_3}v_3. \qquad (2.55)$$

We recall that in dynamic elasticity we used expressions of the form $\partial^2 u/\partial t^2$ for the components of acceleration. In effect, we considered the terms $v_j\partial v_i/\partial x_j$ to be small in comparison with $\partial v_i/\partial t$, and thereby neglected the former. In fact, we implicitly introduced an additional assumption that not only are the $\partial u_i/\partial x_j$ small in comparison with unity, but their partial derivatives with respect to t are small as well. This is common in textbooks on elasticity; however, the latter assumption does not follow from the first, because smallness of a function does not imply smallness of its derivative.

In the next section, we will apply (2.55) to a model of hydrodynamics.

2.22 The Equations of Hydrodynamics

We would like to formulate the equations that govern the motion of a liquid. As we have indicated, we shall use the Eulerian description, but we must combine this with the Lagrangian viewpoint in order to properly express time rates of change following the motion. We begin with the equilibrium equations (2.23), shown previously to hold for any kind of medium:

$$\frac{\partial \boldsymbol{\sigma}_1}{\partial x_1} + \frac{\partial \boldsymbol{\sigma}_2}{\partial x_2} + \frac{\partial \boldsymbol{\sigma}_3}{\partial x_3} + \mathbf{F} = 0.$$

Recall that $\mathbf{F}$ was a force per unit volume. In hydrodynamics, the density of a liquid can change significantly, and the external forces are predominantly gravitational. It is therefore convenient to express the external force densities on a normalized basis. So we will change $\mathbf{F}$ in the above to $\rho\mathbf{F}$ and reinterpret $\mathbf{F}$ as the force per unit mass at a point. Making use of d'Alembert's principle, by which the inertial force density $-\rho D\mathbf{v}/Dt$ can be incorporated into the external forces, we get

$$\frac{\partial \boldsymbol{\sigma}_1}{\partial x_1} + \frac{\partial \boldsymbol{\sigma}_2}{\partial x_2} + \frac{\partial \boldsymbol{\sigma}_3}{\partial x_3} = \rho\frac{D\mathbf{v}}{Dt} - \rho\mathbf{F}.$$

We can base the dynamical equation for a liquid on this result. Recall that $\boldsymbol{\sigma}_k$ is a stress vector acting on the elementary surface whose normal is $\mathbf{i}_k$. In an ideal liquid, the stresses at a point do not depend on direction; furthermore, the particular stress we call *pressure* is always normal to any surface on which the liquid impinges. This means that $\boldsymbol{\sigma}_k = -p\mathbf{i}_k$, where p is the pressure in the liquid at the point of interest. (To explain the "minus" sign, we recall that we agreed to call stretching-type stresses positive. For a liquid, we call compression-type stresses positive instead.) Substituting σ_k into the equation, we get

$$-\frac{\partial p}{\partial x_1}\mathbf{i}_1 - \frac{\partial p}{\partial x_2}\mathbf{i}_2 - \frac{\partial p}{\partial x_3}\mathbf{i}_3 = \rho\frac{D\mathbf{v}}{Dt} - \rho\mathbf{F}. \qquad (2.56)$$

Dividing this by ρ and rearranging, we get the component equations

$$\frac{\partial v_1}{\partial t} + v_1\frac{\partial v_1}{\partial x_1} + v_2\frac{\partial v_1}{\partial x_2} + v_3\frac{\partial v_1}{\partial x_3} = F_1 - \frac{1}{\rho}\frac{\partial p}{\partial x_1},$$

$$\frac{\partial v_2}{\partial t} + v_1\frac{\partial v_2}{\partial x_1} + v_2\frac{\partial v_2}{\partial x_2} + v_3\frac{\partial v_2}{\partial x_3} = F_2 - \frac{1}{\rho}\frac{\partial p}{\partial x_2},$$

$$\frac{\partial v_3}{\partial t} + v_1\frac{\partial v_3}{\partial x_1} + v_2\frac{\partial v_3}{\partial x_2} + v_3\frac{\partial v_3}{\partial x_3} = F_3 - \frac{1}{\rho}\frac{\partial p}{\partial x_3}. \qquad (2.57)$$

These are three equations containing the five unknown functions v_k, p, ρ. Experience tells us that we need more equations to complete the setup. In situations where the liquid can be considered incompressible, the equation of incompressibility is added. For a compressible liquid or gas, we need additional (constitutive) relations between p and ρ, obtained from other areas of physics.

The dynamical equations are so important that we would like to offer another derivation. Let us consider an arbitrary volume V of liquid, with boundary surface S, and write out Newton's second law for the mass of fluid contained therein. The motion of this mass is determined by both the external forces and the reaction of the remaining liquid. The resultant external force is

$$\int_V \mathbf{F} \, dm = \int_V \mathbf{F}\rho \, dV.$$

Choosing a surface element with normal $\boldsymbol{\nu}$ in the liquid, we know that a stress "vector" $\boldsymbol{\sigma}_\nu$ acts on the element. We continue to accept the assumption that the pressure is the same in all directions at a point, and is directed along the normal so that $\boldsymbol{\sigma}_\nu = -p\boldsymbol{\nu}$. Therefore, the resultant force due to these stresses is

$$\int_S \boldsymbol{\sigma}_\nu \, dS = - \int_S p\boldsymbol{\nu} \, dS.$$

Finally, the inertia term is

$$\int_V \frac{D\mathbf{v}}{Dt} \, dm = \int_V \frac{D\mathbf{v}}{Dt}\rho \, dV.$$

Newton's second law applied to the volume V (with use of the solidification principle) gives

$$\int_V \frac{D\mathbf{v}}{Dt}\rho \, dV = \int_V \mathbf{F}\rho \, dV - \int_S p\boldsymbol{\nu} \, dS.$$

Let us transform the last integral term using the formula

$$\int_V \frac{\partial f}{\partial x_k} g \, dV = \int_S f g n_k \, dS - \int_V \frac{\partial g}{\partial x_k} f \, dV,$$

for integration by parts. This term has components $\int_S p n_k \, dS$, where the n_k are the components of $\boldsymbol{\nu}$. We write

$$\int_S p n_k \, dS = \int_V \frac{\partial p}{\partial x_k} \, dV,$$

and therefore we have

$$\int_S p\boldsymbol{\nu} \, dS = \int_V \left(\frac{\partial p}{\partial x_1}\mathbf{i}_1 + \frac{\partial p}{\partial x_2}\mathbf{i}_2 + \frac{\partial p}{\partial x_3}\mathbf{i}_3 \right) dV.$$

Collecting, we get

$$\int_V \frac{D\mathbf{v}}{Dt} \rho \, dV = \int_V \mathbf{F} \rho \, dV - \int_V \left(\frac{\partial p}{\partial x_1} \mathbf{i}_1 + \frac{\partial p}{\partial x_2} \mathbf{i}_2 + \frac{\partial p}{\partial x_3} \mathbf{i}_3 \right) dV,$$

or

$$\int_V \left[\frac{D\mathbf{v}}{Dt} \rho - \mathbf{F} \rho + \left(\frac{\partial p}{\partial x_1} \mathbf{i}_1 + \frac{\partial p}{\partial x_2} \mathbf{i}_2 + \frac{\partial p}{\partial x_3} \mathbf{i}_3 \right) \right] dV = \mathbf{0}.$$

Because V is arbitrary, we have

$$\frac{D\mathbf{v}}{Dt} \rho - \mathbf{F} \rho + \left(\frac{\partial p}{\partial x_1} \mathbf{i}_1 + \frac{\partial p}{\partial x_2} \mathbf{i}_2 + \frac{\partial p}{\partial x_3} \mathbf{i}_3 \right) = \mathbf{0}.$$

(Why? Use arguments involving continuity.) This is equation (2.56).

2.23 D'Alembert–Euler Equation of Continuity

It is obvious that the density ρ is related to the velocity field $\mathbf{v}(\mathbf{x})$. Let us find this relationship, which is called the *equation of continuity*. We consider a fluid with the property that any identifiable fluid volume (i.e., one that always consists of the same particles) always has the same mass. Thus, we trace some elementary connected volume, writing its mass m as $m = \rho^* V$, where ρ^* is the average density of fluid within the volume V. Because m stays constant during the motion, we get

$$\frac{D}{Dt}(\rho^* V) = 0,$$

where D/Dt is the total derivative. Differentiating, we get

$$\frac{D\rho^*}{Dt} V + \rho^* \frac{DV}{Dt} = 0. \tag{2.58}$$

Let us find the expression for DV/Dt. Recall that in Section 2.11 we found the relative change θ of an elementary volume through the components of the strain tensor:

$$\theta = \varepsilon_{11} + \varepsilon_{22} + \varepsilon_{33}.$$

Also recall that for an arbitrary deformation at any point there are "principal directions." Along these three mutually perpendicular directions, the elementary portion of material is stretched or compressed only, so there are no shear deformations. If we erect a Cartesian frame with coordinate axes along the principal directions at a given point, then the sides Δx_i of an elementary right parallelepiped become, after deformation, $\Delta x_i(1 + \partial u_i/\partial x_i) =$

$\Delta x_i(1+\varepsilon_{ii})$, up to second-order quantities. Calculating the change of volume of the material parallelepiped from its original volume $V = \Delta x_1 \Delta x_2 \Delta x_3$, we get

$$\Delta V = (\varepsilon_{11} + \varepsilon_{22} + \varepsilon_{33})\Delta x_1 \Delta x_2 \Delta x_3 = \theta \, \Delta x_1 \Delta x_2 \Delta x_3,$$

up to small quantities. We remember that θ is one of the invariants of the strain tensor, hence we can calculate it for any Cartesian frame using the same formulas. Let us calculate this in terms of the velocities v_i. If we consider a material volume V at time t and at the next time instant $t + dt$ with dt infinitesimally small, we find that the field of infinitesimally small displacements (referenced to the state at time t) of the fluid volume is given as $\mathbf{v}(\mathbf{x}, t)\, dt$. Hence the corresponding strain components are

$$\varepsilon_{11} = \frac{\partial v_1}{\partial x_1}\, dt, \qquad \varepsilon_{22} = \frac{\partial v_2}{\partial x_2}\, dt, \qquad \varepsilon_{33} = \frac{\partial v_3}{\partial x_3}\, dt,$$

so up to small quantities, we have

$$\Delta V = \left(\frac{\partial v_1}{\partial x_1} + \frac{\partial v_2}{\partial x_2} + \frac{\partial v_3}{\partial x_3} \right) V \, dt.$$

This increment ΔV is the one that participates in the definition of the total derivative DV/Dt, so up to small quantities,

$$\frac{DV}{Dt} = \left(\frac{\partial v_1}{\partial x_1} + \frac{\partial v_2}{\partial x_2} + \frac{\partial v_3}{\partial x_3} \right) V.$$

This, when substituted into (2.58), gives us

$$\frac{D\rho^*}{Dt} + \rho^* \left(\frac{\partial v_1}{\partial x_1} + \frac{\partial v_2}{\partial x_2} + \frac{\partial v_3}{\partial x_3} \right) = 0,$$

after division through by V. Letting V shrink proportionally to zero size, we can replace the average density ρ^* by the actual density ρ at point $\mathbf{x}$. We finally have the equation of continuity:

$$\frac{\partial \rho}{\partial t} + v_1 \frac{\partial \rho}{\partial x_1} + v_2 \frac{\partial \rho}{\partial x_2} + v_3 \frac{\partial \rho}{\partial x_3} + \rho \left(\frac{\partial v_1}{\partial x_1} + \frac{\partial v_2}{\partial x_2} + \frac{\partial v_3}{\partial x_3} \right) = 0. \qquad (2.59)$$

This can be written alternatively as

$$\frac{\partial \rho}{\partial t} + \frac{\partial (\rho v_1)}{\partial x_1} + \frac{\partial (\rho v_2)}{\partial x_2} + \frac{\partial (\rho v_3)}{\partial x_3} = 0,$$

which looks like

$$\frac{\partial \rho}{\partial t} + \operatorname{div}(\rho \mathbf{v}) = 0 \qquad (2.60)$$

in the compact symbolism of vector analysis.

Note that for an incompressible liquid we have $D\rho/Dt = 0$, and thus the condition of incompressibility is

$$\frac{\partial v_1}{\partial x_1} + \frac{\partial v_2}{\partial x_2} + \frac{\partial v_3}{\partial x_3} = 0. \tag{2.61}$$

This is also written as

$$\text{div } \mathbf{v} = 0.$$

To complete the system of equations for such a liquid, it is enough to supplement the dynamical equations with the incompressibility equation. For compressible liquids and gases, we must usually add a constitutive relation between p and ρ:

$$p = p(\rho).$$

Relations of this type are derived in physics; they play the same role in hydrodynamics that Hooke's law plays in elasticity. To complete the setup, we need boundary conditions, say,

$$\mathbf{v}\big|_S = \mathbf{v}_0 \qquad \text{for } t > 0,$$

and initial conditions:

$$\mathbf{v}\big|_{t=0} = \mathbf{v}^* \qquad \text{for } \mathbf{x} \in V.$$

2.24 Some Other Models of Hydrodynamics

The equations describing an ideal homogeneous fluid can be extended to form models applicable in other situations. One of these involves a medium consisting of a mixture of several liquids or gases. Such models could be also applied to a multiphase (liquid–gas) state of a medium. These models are approximate, and extensive theoretical development has been motivated by the requirements of myriad applications.

An important version of hydrodynamics is geared toward the dynamics of viscous liquids. When a solid body moves through a liquid, the liquid resists the motion. Experiments show that this "drag force" is, to a first approximation, proportional to the speed of the body. Although the shape of the body plays an essential role in the computation of drag, the tendency of the liquid itself to resist motion has a name: *viscosity*. We previously discussed the case of an ideal liquid, where the stress is purely normal. In a viscous liquid, nonzero tangential stresses exist and must be taken into account. The simplest model for this is referred to as Newton's law for rectilinear, laminar flow. It can be written as

$$\sigma_{21} = \mu \frac{\partial v}{\partial n},$$

where σ_{21} is the tangential stress on the surface perpendicular to the flow direction, μ is the *coefficient of viscosity*, and the derivative $\partial v/\partial n$ is taken along the direction normal to the flow. We see that the velocity components are involved in the constitutive law for a viscous liquid. To characterize the gradients in a nonhomogeneous velocity field, we can utilize a tensor $\{\dot{s}_{ij}\}$ of velocities of deformation. This resembles the strain tensor for small deformations:

$$\dot{s}_{ij} = \frac{1}{2}\left(\frac{\partial v_i}{\partial x_j} + \frac{\partial v_j}{\partial x_i}\right).$$

The best known case is the set of equations of motion for a viscous incompressible liquid, for which the constitutive law is taken in the form

$$\sigma_{ij} = 2\mu\dot{s}_{ij} - p\,\delta_{ij}.$$

Here, δ_{ij} is the Kronecker symbol defined on page 114. For an ideal liquid, only the term $-p\delta_{ij}$ is taken into account. Repeating the way in which the equations for an ideal liquid were derived, we can arrive at the *Navier–Stokes equations*:

$$\frac{\partial v_i}{\partial t} + v_1\frac{\partial v_i}{\partial x_1} + v_2\frac{\partial v_i}{\partial x_2} + v_3\frac{\partial v_i}{\partial x_3} = F_i - \frac{1}{\rho}\frac{\partial p}{\partial x_i} + \mu\nabla^2 v_i \qquad (i = 1, 2, 3).$$

These are supplemented with the equation of incompressibility

$$\frac{\partial v_1}{\partial x_1} + \frac{\partial v_2}{\partial x_2} + \frac{\partial v_3}{\partial x_3} = 0,$$

along with initial and boundary conditions. The corresponding boundary value problems have attracted much attention in the technical literature because of their wide applicability. They continue to provide mathematicians with challenging problems as well.

2.25 Equilibrium of an Ideal Incompressible Liquid

The equilibrium of a liquid is an important particular case that we see all around us. We will consider a few practical problems using the equilibrium equations for an ideal incompressible liquid, which follow from the dynamic equations when $Dv_i/Dt = 0$:

$$\rho F_1 = \frac{\partial p}{\partial x_1}, \qquad \rho F_2 = \frac{\partial p}{\partial x_2}, \qquad \rho F_3 = \frac{\partial p}{\partial x_3}.$$

First, we solve a hydrostatic problem. The most important class of external forces consists of the potential forces, which can be expressed through a potential function $\mathcal{P}$ as

$$F_i = -\frac{\partial \mathcal{P}}{\partial x_i} \qquad (i = 1, 2, 3).$$

In the field of potential forces, the equilibrium equations become

$$-\rho \frac{\partial \mathcal{P}}{\partial x_i} = \frac{\partial p}{\partial x_i} \qquad (i = 1, 2, 3),$$

and for an incompressible liquid, ρ is constant:

$$\frac{\partial (p + \rho \mathcal{P})}{\partial x_i} = 0 \qquad (i = 1, 2, 3).$$

It follows that

$$p + \rho \mathcal{P} = \text{constant}. \qquad (2.62)$$

This simple equation allows us to find the condition under which two non-mixing incompressible ideal liquids will be in equilibrium in a homogeneous field of potential forces. Indeed, let the densities of the liquids be ρ_1 and ρ_2, respectively. The pressure at each point of the interface between the liquids is the same for each liquid, so both p and $\mathcal{P}$ are continuous across the interface. If we take any direction s tangential to the interface, we get

$$\frac{\partial (p + \rho_1 \mathcal{P})}{\partial s} = 0, \qquad \frac{\partial (p + \rho_2 \mathcal{P})}{\partial s} = 0,$$

so that upon subtraction,

$$(\rho_1 - \rho_2) \frac{\partial \mathcal{P}}{\partial s} = 0.$$

Therefore, on the interface,

$$\mathcal{P} = \text{constant}, \qquad (2.63)$$

which states that the interface is an *equipotential surface*.

We live in a gravitational field that is roughly homogeneous and uniform when considered locally. Taking the $x_3 = z$ axis to be directed vertically downward, we can express the gravitational potential as

$$\mathcal{P} = -gz,$$

where $g \approx 9.8 \text{ m/s}^2$. Hence the equipotential surfaces are horizontal. The fact that the surface of a calm lake takes this form serves to confirm the applicability of the liquid model to everyday effects. At the water surface, the pressure p must equal the atmospheric pressure p_0 (i.e., that of the overlying air), so, from (2.62), the pressure at depth z is given by

$$p - \rho g z = p_0.$$

Simple calculations show that the pressure increases by approximately one atmosphere for each ten-meter increase in depth.

These formulas can be used to give a formal demonstration of Archimedes' principle. We leave this to the reader. We could also show that, for a compressible liquid, the equipotential surfaces correspond to the surfaces of constant pressure and the surfaces of constant density.

Equilibrium of a Rotating Incompressible Liquid

Under internal gravity forces, a volume of liquid in free space takes the form of a ball. This has been demonstrated directly through experiments done in space stations, and can be verified at home by placing a drop of oil into a glass of water. It also partially explains why the Sun is nearly spherical. The ball-like form taken by a mass of liquid in equilibrium is, from an energetic viewpoint, the most economical form possible. Newton's law of gravitation for a homogeneous liquid ball is, in terms of the potential function,

$$\mathcal{P} = -\frac{\kappa}{R}, \qquad R = \sqrt{x_1^2 + x_2^2 + x_3^2},$$

where the constant κ depends on the mass of the liquid and incorporates the universal gravitational constant. Note that the distance R appears to the first power in the denominator of the potential, hence it appears to the second power in the denominator of the force expression. The equipotential surfaces are spheres.

Now, suppose the liquid ball rotates with angular frequency ω about the z-axis ($z = x_3$). What happens to the free boundary? Although this is a problem of steady flow rather than of equilibrium, we can formally consider the liquid in a frame that rotates along with it. In this frame, the liquid appears to be in equilibrium under two force fields: one of these is the gravity field with potential as given above; the other is the field of centrifugal inertia forces, for which the potential can be written as

$$\mathcal{P}_c = -\frac{1}{2}\omega^2 r^2, \qquad r^2 = x_1^2 + x_2^2.$$

By what was done above, the equation for the rotating liquid becomes

$$p - \rho\frac{\kappa}{R} - \frac{1}{2}\rho\omega^2 r^2 = \text{constant}.$$

Note that this is approximate, because we neglect the additional gravitational effects that would be associated with the change in shape of the liquid.

On the free surface of the liquid, where the pressure field is constant, we get

$$\frac{\kappa}{R} - \frac{1}{2}\omega^2 r^2 = \text{constant}.$$

An older theory on the Earth's origin held that the Earth is a solidified mass. It is therefore of interest to evaluate how much the Earth's surface has become nonspherical because of rotation. We naturally choose the z-axis passing through the North and South poles. On the North pole, we have $r = 0$, so denoting $g_0 = \kappa/R_0^2$, with R_0 equal to the distance from the Earth's center to its surface at the pole and g_0 the free-fall acceleration constant, we have

$$\frac{g_0 R_0^{\,2}}{R} + \frac{\omega^2 r^2}{2} = \frac{g_0 R_0^{\,2}}{R_0}.$$

The radii R and r are interrelated through the polar angle θ as $R = r\sin\theta$, so we get

$$\frac{g_0 R_0^{\,2}}{R} + \frac{\omega^2 R^2 \sin^2\theta}{2} = g_0 R_0.$$

From this, it can be calculated that the minimum "radius" is the value R_0 at the pole, and the maximum occurs at the equator. Hence the fractional compression of the sphere corresponding to a "liquid" Earth is

$$\varepsilon = \frac{R_{\text{equator}} - R_0}{R_0} \approx \frac{1}{2}\frac{\omega^2 R_0^2}{g_0} \approx \frac{1}{600}.$$

Geodesic measurements show twice as much compression. The discrepancy might be explained by nonuniformities in solidification, if indeed the solidification hypothesis is applicable.

2.26 Force on an Obstacle

We have all felt a strong wind, or the drag produced by water moving past our bodies as we stand quietly in a river. The problem of how wind can affect buildings (sometimes tearing away roofs) is important, so we wish to understand how to calculate the corresponding forces. We will do this on an elementary level, without using the general equations of hydromechanics. However, the computation of forces on obstacles is one of the central problems of that subject. For various models of a liquid and various assumptions, extensive calculations can be carried out to predict the lift on an airplane wing, the drag force on a moving car, and so on. These problems are difficult, and normally the setup must include, in addition to all the main equations, certain assumptions regarding the behavior of the flow pattern.

One of the first problems to be considered was the computation of the force on an infinite circular cylinder immersed in the laminar flow of an ideal liquid. The flow was taken to be steady at some constant velocity $\mathbf{v}$, and it was assumed that the flow picture would be the same for all cross sections of the cylinder. So the problem was two-dimensional and could be described in terms of two coordinate variables x_1, x_2. D'Alembert himself found that a drag force of zero was predicted, and his surprising result came to be called *d'Alembert's paradox*. The obvious contradiction with our everyday experience can be resolved when we realize that it is impossible to construct a truly infinite cylinder and, moreover, there is viscosity. Any real body of liquid will have boundaries that introduce other phenomena into consideration: the flow will be impeded at any fixed boundary, and waves can appear at a free surface. Waves and vortices can also appear internal to the flow, and these can have serious implications regarding the computation of drag force.

To get results closer to the real ones, let us construct a simple model of how airflow acts on a person standing in the street. We will take the ambient wind to be directed horizontally at speed v, and the cross sectional area of the person's body to be S. Let us suppose that where the wind meets its obstacle, the flow stream changes direction by an abrupt 90°. We will neglect the influence of this portion of the air on the motion of the rest of the air. So our assumptions are pretty rough, but they will permit us to carry through with a basic calculation.

Let us use Newton's second law in integral form for the direction along the wind stream. The impulse of the drag force F during a time length t under steady conditions is equal to Ft. During the same time interval, a cylindrical slab of air of length vt and cross sectional area S, moving at velocity v, is halted by the obstacle. The mass of this slab is $m = \rho vtS$. So the change in linear momentum of the moving slab of air is

$$mv = \rho v^2 tS.$$

By Newton's second law, we have

$$Ft = \rho v^2 tS,$$

and thus

$$F = \rho Sv^2. \tag{2.64}$$

Although this formula is rooted in simple assumptions, it provides a good understanding of drag force. The first thing to note is the dependence of F on the square of the velocity v. The same dependence comes out of more complex models. A similar statement can be made regarding the dependences of F on ρ and S. A formula due to Rayleigh[13] gives the force on an infinite plate of width a standing orthogonally to a flow stream, assuming the stream becomes stagnant behind the obstacle. The force on height h of

---

[13] John William Strutt (Lord Rayleigh) (1842–1919).

the plate is given by

$$F_R = \rho v^2 ah \frac{2\pi}{\pi + 4}. \tag{2.65}$$

Taking a rectangular portion of area $S = ah$ and comparing with (2.64), we see that the ratio is

$$\frac{F_R}{F} = \frac{2\pi}{\pi + 4} \approx 0.88.$$

So our comparatively simple model gave us a respectable approximation to the better result (2.65). Now, Rayleigh's formula is also approximate because it assumes the existence of a stagnation line (along which the fluid velocity is discontinuous); it also neglects boundary effects and other things that can turn out to be important. But our formula does display the true dependence of F on the main parameters of the problem. More accurate calculations require a numerical approach via the finite element method, by which one can take into account many factors impacting the real motion.

Let us apply (2.64) and obtain a numerical estimate for a fictitious human being. The density of air is about 1.3 kg/m³, and the area S should be around 0.5 m × 2 m = 1 m²:

$$F = 1.3\, v^2 \text{ N.}$$

So, for $v = 20$ m/s², we get $F = 520$ N, whereas for $v = 30$ m/s², we get $F = 1170$ N. The actual force will be less because the shadow region behind the body is smaller than we assumed. But the estimate allows us to see why a storm wind of 20 m/s can stop you in your tracks. A 30 m/s wind can drag you along with it, and anything faster than that can sweep you off your feet.

Chapter Three

Elements of the Strength of Materials

3.1 What Are the Problems of the Strength of Materials?

Everyday experience has made us familiar with the properties of many things. Without performing any calculations we can evaluate whether we can bend or break a tree branch. We expect a dinner plate to shatter when dropped on a hard floor. We know that a tiny nail cannot hold up a huge shelf full of books. We are aware that boiling water should not be poured into a cool crystal glass.

Most of our understanding is, however, qualitative in nature. Now, there is nothing inherently wrong with this; a good design engineer will depend on a refined level of intuition for much of what he or she can accomplish. But there are circumstances under which common sense can tell us little or, worse yet, yield a substantially wrong picture of things. Such is the case for structures exposed to atomic radiation, and for those made of artificial composite materials having unusual properties. It happens with ordinary things subjected to unusual environmental factors. In these cases, an engineer needs to understand more precisely the processes involved. Failure of certain structures (bridges, buildings, ships, airplanes, nuclear reactors) can put our lives at risk; furthermore, such failures occur all too often — and sometimes for seemingly negligible reasons. Even when danger is not involved, it is of value to know the expected service lifetime of various pieces of equipment. This forces engineers to apply design tools more precise than semi-intuitive considerations. Sometimes performance predictions can be based on direct experimentation, but often this involves an impractical amount of work. The usefulness of numerical calculations done for mathematical models of objects and processes is evident.

Many people are under the impression that scientists can calculate anything of interest using modern computers. This is far from reality. Some processes can be investigated through computational means, but for others the results sought are completely obscured by numerical error. There are other important sources of error as well. One is the difficulty we sometimes have in constructing an appropriate model of a process or object. The models of solid mechanics are based on the continuum hypothesis, but we also seek to predict effects relating to material behavior on a microscopic level (damage, cracks, etc.). In addition, the models of solid mechanics are accurate only in an average sense; any prediction made for an individual object can be put forth only with some probability. We should add that an engineer cannot predict all environmental factors to which a structure might be subjected. Finally, even in the framework of simple, "strictly accurate" models, the calculations are quite approximate. These are only a few of the reasons why predictions done with computers can fail.

We would like to guarantee some working lifetime for any item we design. From an engineering standpoint, however, we can only produce some numbers that serve as indicators of continued performance over a certain warranty period. For this, we must know the strength properties of those materials out of which the structure is composed, and how the structure functions. We must also have a model to characterize workability. For some structures, this simply means that they will not collapse, while for others it implies restrictions on the displacements of certain elements or changes of shape. Experience in the use of structures suggests that damage can be avoided if we can suitably restrict the stress levels reached in certain structural units. Thus, a principal problem of engineering design is the calculation of strain in these units under extreme conditions of use. An engineer must know the stress and strain values in each structural member and make sure these remain suitably bounded. A century ago this was done on the basis of a science called the *strength of materials*, which used simple linear elastic models of stretched and bent beams, among others. Two- and three-dimensional elastic bodies were then modeled, as were plates and shells; all of these objects belong to a more mathematical science called the *theory of elasticity*. More complex models of plasticity, elastoplasticity, viscoelasticity, and so forth, arose as parts of solid mechanics. These can all be regarded as belonging to the strength of materials. Nonlinearity in constitutive laws, or the geometrical description of strains, brought new effects regarding the stability of structures. Dynamical considerations have brought even more difficulties for the designer. Finally, for many structures, it is impossible to deal only with elastic or nonelastic behavior; rather, we must bring in considerations involving interaction with a surrounding fluid, temperature effects, electromagnetic effects, and so on. These effects were originally relegated to different realms of science and were treated as though they were independent. Now, computerization has brought these effects back together and we have more complex sciences as a result: one of these is the abovementioned strength of materials, which studies the behavior of various materials and structures in complex situations.

3.2 Hooke's Law Revisited

Once people begin using any model extensively, there is an unfortunate tendency toward identification of the model with the real object it is meant to describe. But any mathematical model is only approximate. A model for a more or less complex object, such as any object of mechanics, consists of parts having different natures: some parts are precise (at least within the framework of certain chosen assumptions), while others are taken directly

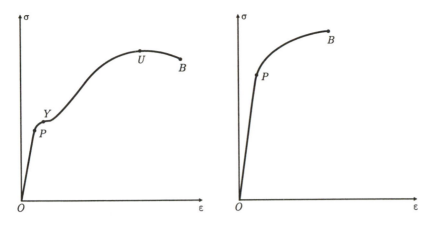

Figure 3.1 Examples of stress–strain curves.

from experiment so that their accuracy is determined by measurement methods and the supplementary theories on which those methods were based. For
models based on experimental data, workers try to avoid additional inaccuracy by incorporating theoretical considerations into the main equations with
as much accuracy as possible. Let us see how this is done for the model of
a linearly elastic body.

Hooke's law is always based upon the results of tests. The equations of
equilibrium and expressions for strains are not absolutely precise because
they exploit hypotheses on the smallness of deformation; however, they are
precise if those assumptions are taken as absolute axioms. How do the precise
equations and the approximate Hooke's law collaborate? Experience shows
that this combination can be successful in predicting solutions to complex
problems. We would like to understand how and why this accuracy of the
final model arises, in view of the fact that Hooke's law is not extremely accurate. In particular, Hooke's law gives the relation between stresses and
strains, whereas the derivatives of the strains appear in the main equations.
(The reader is aware that smallness of a function does not guarantee smallness — or even boundedness — of its derivatives.) Let us begin with a
discussion of how Hooke's law is found in practice today.

Every mechanical engineering student has used a test apparatus to plot
the stress–strain curves for several materials. The experiment is done with
standard specimens under tension. We present two representative plots in
Figure 3.1. The plot on the left is typical for medium-carbon structural steel,
whereas that on the right is typical for hard steels and some nonferrous alloys.
These curves are reasonably representative, but shapes for other materials
can vary greatly. For many plastic materials, we cannot even draw such
a diagram because for those materials, the stresses will exhibit a temporal
dependence.

Steel is an important material in practice. Let us discuss the plot on the

Figure 3.2 Geometry of a steel test specimen.

left in more detail. The specimen is shaped as indicated in Figure 3.2, with a cylindrical test portion (having circular cross section) in the middle. Its relative elongation $\varepsilon = \Delta L/L$ can be measured by several methods that vary in accuracy. On the vertical axis (Figure 3.1) we show the cross sectional stress σ calculated as the force applied to the specimen divided by the initial cross sectional area. Portions of our diagram have been exaggerated to afford us a better look at certain characteristic features.

The first thing to notice is the linear segment OP. The ordinate of point P is called the *proportional limit*; the strain value at P is that below which we can apply Hooke's law,

$$\sigma = E\varepsilon.$$

Slightly above P, there is another point whose ordinate is the *elastic limit*. Until we reach this stress, relaxing the tension to zero results in the stresses and strains returning along the same curve. When the stress exceeds this value, we find that relaxing the tension to zero leaves a residual strain related to the phenomenon of plasticity. It is common to refer to the portion of the diagram between the origin and the proportional limit as the *elastic region* (despite the fact that the actual elastic region is a bit bigger). The rest of the curve is called the *region of plasticity*. In this region, we select a point Y, called the *yield point*, where the curve becomes almost horizontal; this can be interpreted as a point about which we can increase the strain without increasing the stress. The maximum ordinate of the curve is called the *ultimate*, or *tensile, strength* of the material. The final point B is the *breaking strength*.

We are interested in the linear (elastic) part of the diagram, but let us take a moment to discuss the rather strange form of the curve. The diagram shows that specimen rupture (point B) happens at a tension *below* the ultimate strength, which seems a bit nonintuitive to those who have not observed the process happen to a real specimen. The explanation is that an additional process occurs during extension: a reduction in cross sectional area. This reduction is first characterized by Poisson's ratio in the elastic region. In the diagram, stress is expressed as the tensile force divided by the *initial* cross sectional area. But because this area shrinks, the actual stress exceeds that shown in the diagram. At some point, the increasing stress begins to produce one to several "bottlenecks" on the specimen. In these regions, both elongation and reduction in cross sectional area are considerable, and the actual stress (calculated using the actual cross sectional area) is larger than that shown in the diagram. An explanation for this strange behavior in the plastic region must be based on microscopic considerations; we return to the issue later on.

Now we would like to discuss the elastic part of the diagram and the accuracy of linear approximation. We have said many times that no model of a real object can be absolutely exact; all models are approximate. So the linearity of Hooke's law should be only a good approximation. This is actually the case. Let us consider the experimental conditions under which we may obtain Hooke's law.

First, it is impossible to guarantee a fully homogeneous state in the cross section of a specimen under tension. Indeed, we must attach the specimen to the test machine, and we cannot guarantee uniformity of the attachment. We can only suppose that, far from the attachment regions, the state is close to homogeneous, but we cannot really tell what happens inside the specimen without solving the corresponding boundary value problem.

Next is the problem of accuracy in the measurement of strains and stresses. With regard to both of these the accuracy is restricted. The next point concerns the model itself. Here, we calculate the stress using the initial area of the cross section. But the linear dimensions of the cross section experience Poisson's effect and, hence, the reduction in area is (at least) quadratic with respect to the longitudinal strain. This means that nonlinearity necessarily enters into the law.

Finally, there is the problem of temperature change. Any strain in a body implies a change in temperature (the reader may have learned this while bending a piece of wire in an attempt to break it) and conversely. Moreover, temperature-related (heat transfer) processes take time to occur. This means that different rates of increase of tension can result in different stress–strain plots. Time dependence can also bring nonlinearities into a process that would otherwise be linear. This shows that the early development of a theory can often proceed better *without* the benefit of extremely precise measurements. We can only imagine what would have happened to the strength of materials if Hooke's law had been formulated as a nonlinear law that displayed dependences on both time and temperature. (The result is a complex theory under which it is hard to obtain analytical solutions.)

However, all the correction terms required to bring Hooke's law into line with reality bring only small perturbations. (The perturbations are not infinitely small in a mathematical sense, but are small in a practical sense. This means that, through simple calculation and comparison with experiment, we find that omission of some terms gives an acceptable relative error.) These perturbations in turn produce relatively small errors when inserted into the equations of a boundary value problem. For statics problems, there are theorems on small perturbations for a linear operator equation where the operator has a continuous inverse; these guarantee that the solution is close to that of the unperturbed equation. Some of these theorems are based on

Banach's[1] contraction mapping principle, while others are based on series expansion of the operator (assuming that the perturbations can be similarly expanded in power series with respect to some parameter of the problem).

The contraction mapping principle is one of the most useful tools in all of mathematics, and now that we have mentioned it, we should offer a brief sketch for the reader. The principle helps us solve equations of the general form

$$x = Ax, \tag{3.1}$$

where A is a given operator (i.e., a function that maps points of a set into the same set). Suppose, for instance, that A maps vectors of the Euclidean space $\mathbb{R}^n$ into vectors of the same space $\mathbb{R}^n$. Then (3.1) would take the form

$$\mathbf{x} = A\mathbf{x}, \qquad \mathbf{x} = (x_1, \ldots, x_n). \tag{3.2}$$

Note that the vectors $\mathbf{x}$ we seek are those having a certain special property: they are left unmodified by the action of the operator A. Any such vector is called a *fixed point* of A. An important case of (3.2) is the problem having

$$A\mathbf{x} = B\mathbf{x} + \mathbf{c}, \tag{3.3}$$

where B is a linear operator and $\mathbf{c}$ is a constant vector. Then B can be represented by a matrix, and in this case (3.2) looks like

$$\begin{pmatrix} x_1 \\ \vdots \\ x_n \end{pmatrix} = \begin{pmatrix} b_{11} & \cdots & b_{1n} \\ \vdots & \ddots & \vdots \\ b_{n1} & \cdots & b_{nn} \end{pmatrix} \begin{pmatrix} x_1 \\ \vdots \\ x_n \end{pmatrix} + \begin{pmatrix} c_1 \\ \vdots \\ c_n \end{pmatrix}. \tag{3.4}$$

The contraction mapping theorem guarantees both the existence and uniqueness of a fixed point of A, provided that A satisfies certain conditions. One essential condition is phrased in terms of a numerical constant q that serves to characterize the behavior of the operator in a particular sense. For the problem (3.4), a suitable q turns out to be given by

$$q = \max_{1 \leq i \leq n} \sum_{j=1}^{n} |b_{ij}|.$$

The significance of q is as follows. We can perform *iterations* according to the formula

$$\begin{pmatrix} x_1^{(k+1)} \\ \vdots \\ x_n^{(k+1)} \end{pmatrix} = \begin{pmatrix} b_{11} & \cdots & b_{1n} \\ \vdots & \ddots & \vdots \\ b_{n1} & \cdots & b_{nn} \end{pmatrix} \begin{pmatrix} x_1^{(k)} \\ \vdots \\ x_n^{(k)} \end{pmatrix} + \begin{pmatrix} c_1 \\ \vdots \\ c_n \end{pmatrix}. \tag{3.5}$$

---

[1] Stefan Banach (1892–1945).

This means that we choose an initial vector $(x_1^{(0)}, \ldots, x_n^{(0)})$, insert it into the right-hand side of (3.5), and carry out the matrix multiplication to produce a new vector $(x_1^{(1)}, \ldots, x_n^{(1)})$, as shown on the left-hand side. We then insert $(x_1^{(1)}, \ldots, x_n^{(1)})$ into the right-hand side, do the matrix multiplication again, and produce another new vector $(x_1^{(2)}, \ldots, x_n^{(2)})$. These iterations can be carried out ad infinitum, and we could ask whether a limit vector will be approached. The answer is yes, provided that $q < 1$. The effect of having q satisfy this condition is that the vector "iterates" will fall closer and closer together in $\mathbb{R}^n$ as k increases. So the operator A has the effect of decreasing the distance between points, and is said to be a *contraction operator*. Banach's principle states that A will have a unique fixed point, and that the above process of iteration will lead us from *any* initial point (vector) to the fixed point. The smaller the nonnegative value q, the faster this convergence will occur.

Banach's contraction mapping principle also applies to nonlinear operators acting in spaces of infinite dimension, and these are the cases ordinarily encountered in practice. A common choice of starting point for the iterations is the solution of the linearized mechanics problem.

3.3 Objectiveness of Quantities in Mechanics Revisited

In a previous section, we discussed the problem of objectiveness in mechanics. Let us return to it once more. Any entity of Newtonian mechanics will possess certain absolute properties. These can be described in a coordinate frame, but because they are properties of a body, they should not change with a change of frame. Such properties are called *objective*. Of course, when we change the frame and have some relations between the old and new frames, the representation of these properties in the new frame comes about by way of certain strict rules. This is done in the same way for any parameter describing a point of a mechanical body. We have said that, in classical mechanics, each point of a continuum medium can be described using algebraic objects of three kinds: scalars, vectors, and tensors. All the descriptive quantities depend on the point chosen in the continuum body, and this is reflected in a dependence on the coordinates. During a change of frame, the coordinates of points of the body change for all the kinds of parameters, but the behavior of the representative components of the parameters is different. For example, the temperature at a point of a body should not depend on the change of coordinates; it remains the same independently of the change, even if the new frame moves. Parameters of this kind are called *scalar objective functions* defined on the body, and so they relate to algebraic scalars. Force presents us with an example of a vectorial quantity. If a force acts on a body, at any instant, it is "fixed" with respect to the body; therefore, when one rotates the coordinate frame, all the force components in the new frame should be reprojected respectively. The components of all parameters of vector type (e.g., forces, positions, velocities, accelerations) are recalculated in the same manner, using only formulas for the transformation of the frame vectors. There is therefore no need to remember all the particular transformation rules for each kind of vectorial parameter.

We have said that scalars and vectors are particular (and the simplest) cases of the class of tensor parameters, for which the rules of transformation of components can be derived in general form. We recall that the strain and stress tensors are tensors of rank two, whereas vectors and scalars could be called, respectively, tensors of rank one and zero. An objective tensor parameter in continuum mechanics is "affixed" to a point of the body in such a way that it depends on the position with respect to other points of the body, but not on the external frame. In this way, it is similar to a vector parameter related to the same point. However, there is a difference. Let us consider it in more detail using the stress tensor and a force vector.

In a Cartesian frame, we derived a representation formula for the stress vector σ_n acting on the unit elementary plane area with normal $\mathbf{n}$. This was done through three stress vectors $\sigma_1, \sigma_2, \sigma_3$ acting at the same point on the unit elementary plane areas with normals parallel to the coordinate axes x_1, x_2, x_3, respectively:

$$\sigma_n = \sigma_1 n_1 + \sigma_2 n_2 + \sigma_3 n_3.$$

Thus, the components of σ_n are fully defined by the components of the three vectors $\sigma_1, \sigma_2, \sigma_3$; we could therefore introduce an object called the stress tensor, described by the nine components σ_{ij} of these three vectors. We could join the components of the frame vectors to the σ_{ij} and define a nine-dimensional vector $\sigma_{11}, \ldots, \sigma_{33}$. This might lead us to believe that we could bypass the tensor concept and merely extend the vector concept to higher dimensions. This is sometimes done for the solution of certain special problems of mechanics posed in a fixed (say, Cartesian) frame, but it is conceptually wrong.

There are certain rules describing how to transform the components of a vector if we change the frame. If we rotate a frame, for example, then we can recalculate the components of a given force vector in the new frame. Analytic geometry tells us that the new components are linear combinations of the old components with coefficients that are cosines of the angles between the new and old axes. This is unfortunately not quite true for components σ_{ij} of the stress tensor: the form relating the two sets of components is still linear, but the coefficients are products of cosines. So no subset of $\{\sigma_{ij}\}$ can form the components of a vector (or even a vector of nine dimensions), but it is clear that $\{\sigma_{ij}\}$ does characterize an objective property of the state of a body. It is an example of a tensor of the second rank. The strain tensor is also a second-rank tensor, and it shares some properties with the stress tensor because both are symmetric.

In a certain frame, the components of a second-rank tensor are represented by a square matrix. There is a deep correspondence between the properties of square matrices and second-rank tensors; in fact, everything we know about the former can be reformulated in terms of the latter. So we already know a lot about tensors. In particular, the theory of eigenvalues and eigenvectors applies directly to stresses and strains. One consequence is that, at each point of a strained body, there are three mutually orthogonal directions such that the deformation there can be regarded as simple stretching (or compression) in these three directions.

The correspondence between square matrices and tensors is even deeper than we have shown. The transformation formulas for tensor components under a change of frame are exactly those for matrix elements regarded as a linear transformation of space. So we can simply consider matrix representations of second-rank tensors in the same manner as we consider a row or column matrix to represent an objective vector.

The question could be raised whether there are objective quantities other than vectors and tensors of the second rank. The answer is affirmative. If we consider how we must change the elastic coefficients of a material in Hooke's law in order to preserve the objectivity of the stress and strain tensors under frame transformation, we find that the necessary transformation formulas depend on quartic forms composed of the same cosines mentioned above. This naturally leads to the idea of a fourth-rank tensor. Unfortunately, four-dimensional matrices are not typically discussed in elementary courses on analytic geometry. The subject known as multilinear algebra serves to treat tensors of any finite rank.

3.4 Plane Elasticity

Three-dimensional elasticity problems represent a real challenge. Most require numerical methods for their solution. But the few available analytic solutions are of great importance. They bring us an understanding of what happens in similar situations, yield useful analogies, and provide canonical problems for the validation of numerical approaches.

Especially important are the problems of *plane elasticity*. For these, general methods of analytic solution have been based on the theory of functions of a complex variable. A plane elasticity problem can be formulated in terms of the two Cartesian coordinates x_1 and x_2 only. The two subclasses of such problems consist of the (1) plane deformation problems, and (2) plane stress problems.

A problem of plane deformation comes with the following assumptions. The components of the displacement vector of a body $\mathbf{u} = (u_1, u_2, u_3)$ do not depend on x_3, and $u_3 = 0$. This brings us to the relations

$$\varepsilon_{33} = 0, \qquad \varepsilon_{13} = 0, \qquad \varepsilon_{23} = 0,$$

and thus for a homogeneous isotropic body, we have

$$\sigma_{13} = \sigma_{23} = 0.$$

Now $\theta = \partial u_1/\partial x_1 + \partial u_2/\partial x_2$, and Hooke's relations reduce to the following:

$$\sigma_{ii} = \lambda\theta + 2\mu\varepsilon_{ii} \quad (i = 1, 2),$$
$$\sigma_{33} = \lambda\theta,$$
$$\sigma_{12} = 2\mu\varepsilon_{12}.$$

The two nontrivial equilibrium equations are

$$\frac{\partial\sigma_{11}}{\partial x_1} + \frac{\partial\sigma_{12}}{\partial x_2} + \rho F_1 = 0,$$
$$\frac{\partial\sigma_{12}}{\partial x_1} + \frac{\partial\sigma_{22}}{\partial x_2} + \rho F_2 = 0. \tag{3.6}$$

The reader can try to derive the Lamé equations in displacements, and formulate boundary conditions for the problem.

A plane stress problem can arise in one of two ways. For the stress tensor, we can assume

$$\sigma_{33} = 0, \qquad \sigma_{13} = 0, \qquad \sigma_{23} = 0. \tag{3.7}$$

Now the displacement components u_i can depend on x_3. However, using the condition $\sigma_{33} = 0$, we can eliminate ε_{33} from the relations and obtain the following version of Hooke's law:

$$\sigma_{ii} = \lambda^*\theta + 2\mu\varepsilon_{ii} \quad (i = 1, 2),$$
$$\sigma_{12} = 2\mu\varepsilon_{12},$$

where $\lambda^* = (1 - 2\nu)\lambda/(1 - \nu)$. Equations (3.6) remain the same, and only in the Lamé equations is it necessary to change λ to λ^*. Therefore, from the mathematical viewpoint, the two types of problems are equivalent. We may also derive the same plane stress problem in a different way. In the above, we assumed nothing about the shape of the body, but bodies of plate type occur frequently in engineering structures. For these, it is possible to average the characteristics of the body over the thickness and thereby formulate everything in averaged terms. In particular, we can formulate the assumptions (3.7) for the averaged stresses. In this way, we get exactly the same equations as before.

As a rule, mass forces are neglected in this theory. Indeed, their role is minor in comparison with stresses caused by boundary loading (which enter the model through enforcement of the boundary conditions).

We will not go into the theory of plane elasticity. However, let us mention one interesting fact. It turns out that for a simply connected elastic domain loaded by boundary forces only, the stress field does not depend on Poisson's ratio ν. This gives rise to the theory of *photoelasticity*, which allows us to actually see the stress distribution inside a body. For this, we use a model made of a special material. When placed under stress, the body exhibits obvious lines that correspond to certain stress values (the lines are reminiscent of magnetic field lines made visible by iron filings). By counting the lines, we can calculate the real stresses in the body.

Finally, let us mention that in plane elasticity (as well as in the three-dimensional case) we can use curvilinear coordinates. In this case, the equations take another form. In polar coordinates, the equilibrium equations can be written out as

$$\frac{\partial \sigma_r}{\partial r} + \frac{1}{r} \frac{\partial \tau_{r\theta}}{\partial \theta} + \frac{\sigma_r - \sigma_\theta}{r} = 0,$$
$$\frac{1}{r} \frac{\partial \sigma_r}{\partial \theta} + \frac{\partial \tau_{r\theta}}{\partial r} + 2\frac{\tau_{r\theta}}{r} = 0. \tag{3.8}$$

According to historical notation, here we denote by σ_r the normal component of $\boldsymbol{\sigma}$ on the area having normal in the r-direction, by $\tau_{r\theta}$ the tangential component of the tensor on the same elementary area, and by σ_θ the normal component of stress on the elementary area for which the normal is determined by the change of the angle θ. These elementary areas are mutually orthogonal at each point.

Let us consider two particular problems of elasticity that played important roles in the development of engineering.

3.5 Saint-Venant's Principle

Many years ago, Barre de Saint-Venant (1797–1886) wrote a treatise on the twisting of a cylindrical bar of arbitrary shape. Even someone without an engineering mindset can easily picture such twisting, but from a mathematical viewpoint we need to formulate everything explicitly. In particular, we must pose boundary conditions for the bar. When we twist a bar in practice, we normally do not worry about how the twisting moment is applied. However, in the solution of a corresponding boundary value problem, this issue is extremely important; analytic solutions are available only for certain distributions of external load. For a bar of arbitrary cross section, it is difficult to determine the load distribution for which there is an analytic solution, but Saint-Venant proposed a useful principle. This states that the strain state of a twisted bar, far from the points of application of the external moments, depends very weakly on the distribution of the external load; the main dependence is on the resultant moment and force. In other words, suppose we apply two types of external loads with identical moment and force resultants. Then, sufficiently far from the points of application of these loads, the difference between the strain states is practically negligible.

Figure 3.3 Cross section of an apparatus consisting of two long, parallel cylinders
connected by a thin plate.

This statement follows from our everyday experience but does not appear
so evident when we deal with the equations. Of course, it was necessary to
elaborate on the term "sufficiently far" because practitioners needed some
numerical guidelines. As a general rule, if a is the greatest cross sectional
dimension of the bar, then $3a$ (or, safer yet, $5a$) will suffice as the needed
distance.[2]

Saint-Venant's principle was subsequently extended, beyond the twisting
of bars, to general problems of elasticity. So for distances greater than $3a$
to $5a$, where a is the diameter of the area of load application to a massive body, the strain distribution is approximately the same for all loads
having the same resultant moment and force. This principle was long used
in practical calculations. It was also supported by experiments. However,
its mathematical proof was given only recently. The first papers appeared
about 30 years ago, and work on the topic continues today. The principle
has been proved for massive bulk bodies, but exceptions have been found in
other situations. These mostly occur in cases where a body has very flexible parts across which it can be bent using a load whose resultant moment
and force are both zero (see Figure 3.3 for an example). Because of Saint-
Venant's principle and the linearity of many problems, engineers could often
approach complex elasticity problems through the assumption of lumped
forces and couples in the body of interest. In practice, it was possible to
perform acceptable pen-and-paper calculations for problems that, in their
precise formulations, would have been otherwise out of reach.

Let us note that the statics problems of linear elasticity belong to the
class of linear elliptic problems, among which are included boundary value
problems for Laplace's equation describing a membrane. The solutions of
homogeneous elliptic problems for equations with analytic coefficients are
fairly smooth inside the domain. The behavior of the solutions described by
Saint-Venant's principle is determined, in particular, by the elliptic nature
of the equations of elasticity. For solutions of Laplace's equation, there is
a maximum principle: the greatest and least values of the solution are attained on the domain boundary. This means the following. If we somehow

---

[2]In engineering practice, one encounters many such "rules of thumb," and it is interesting that these often involve factors of 3, 5, or 10. The curious preference for these factors
must have its roots in engineering psychology. However, engineers have also been known
to originate much more precise looking "fudge factors." It is not uncommon to find that
mathematicians, working years later with powerful approximation tools, have been able
to put these guesses on a solid footing.

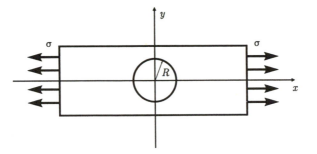

Figure 3.4 The Kirsch problem of plane elasticity.

fix the edge of a membrane that is free of external load, we will see that the deflection of interior points is less than the deflection at the boundary. Moreover, discontinuities in the slope of the membrane edge become smoothed as one moves into the interior of the membrane, where things are described by infinitely differentiable functions. Thus, "bad" behavior of a solution may occur only at the membrane edge. A similar thing happens with solutions of elasticity problems involving no volume forces, and the situation is described by the next problem, stress concentration.

3.6 Stress Concentration

We know that the level of stress determines the strength of real-life structures, and that it is important to understand how the strains distribute themselves inside the structure and at which points the structure is weakest. We have said that some simple problems of elasticity demonstrate behaviors typical of the solutions to complex problems. One useful problem to examine is that of an infinite plate with a circular hole, stretched by forces applied at "infinity." The solution demonstrates the weakness of our intuition about when a steel structure is likely to fail. So consider Figure 3.4. The plate is subjected to stretching along the x-axis, by forces that produce the stress component $\sigma_{11} = \sigma$ at points far removed from the hole. No forces are applied at the hole boundary. This so-called Kirsch problem can be treated in the framework of plane elasticity. We have therefore introduced Cartesian coordinates x, y, and we shall write $\sigma_{11} = \sigma_{xx}$.

Suppose we have drilled only a tiny hole. Can anything serious happen? Intuition might suggest that a structural problem could arise only for a large hole. An exact solution to the Kirsch problem demonstrates otherwise. We shall examine the final answer only, leaving the interested reader to examine the solution process in any textbook on elasticity. On the line $x = 0$, the

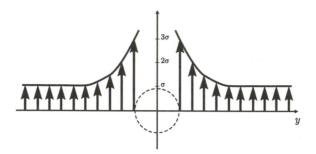

Figure 3.5 Stress concentration near a circular hole.

stress component σ_{xx} is given by the equation

$$\sigma_{xx} = \sigma \left(1 + \frac{1}{2}\frac{R^2}{y^2} + \frac{3}{2}\frac{R^4}{y^4} \right). \tag{3.9}$$

At points far from the hole, σ_{xx} is close to σ (i.e., the stress value for the plate without the hole). But at the edge of the hole, where $y = R$, we have $\sigma_{xx} = 3\sigma$, so σ_{xx} triples near the opening. This kind of behavior of the stress near a special point is called *stress concentration*. The maximal value 3σ is denoted $(\sigma_{xx})_{\max}$. The number 3 is called the *concentration factor*. See Figure 3.5.

The value $(\sigma_{xx})_{\max}$ obviously determines whether the plate will fail under a given applied stress. Observe that the concentration factor is independent of R. This means that a hole of any size can seriously reduce the integrity of the plate.

The concentration factor can be even greater for a rectangular plate of some finite width L; it is approximately $3L/(L-R)$. A more general elliptical hole can also be considered. For an ellipse having axes of length $2a$ and $2b$ along the x and y directions, respectively, the result for an infinite plate is

$$(\sigma_{xx})_{\max} = \sigma \left(1 + \frac{2a}{b} \right). \tag{3.10}$$

We see that the concentration coefficient can far exceed 3 when $a/b \gg 1$; this corresponds to a model of a crack. It seems that, for any stretching force, we could choose the ratio a/b in such a way that the plate will snap in half. However, experience shows that this is not the case. Indeed, the elastic model ceases to apply at some level of stress, and a model from the theory of plasticity must be applied instead.

Holes of other shapes can bring even worse effects. We know we can easily tear a piece of cloth after introducing a slight rip at the edge. This is another example of stress concentration. But in many instances the phenomenon turns out to be harmful rather than useful. For example, a massive ship that looks strong can be loaded with cargo in such a way that extremely high stresses are induced in the deck; a hole in the deck can bring disastrous results. Strangely, stress concentration arises not only where we remove part of a body (e.g., drill a hole), but also at points of inhomogeneity — even if we replace part of the body with material *stronger* than the original. The introduction of additional members can also weaken a structure if done without proper understanding. Indeed, reinforcement of one portion of a structure can come at the expense of the strength of other parts, resulting in a sometimes dangerous weakening of the structure as a whole. This has been known to happen, for example, when the owners of trucks have added extra structural ribs in an attempt to reinforce their cargo area.

Stress concentration can play an extremely important role in structural integrity, and we should be aware that attempts to drill holes or attach additional parts to a structure may bring catastrophic results.

3.7 Linearity vs. Nonlinearity

We know that real bodies consist of atoms. We are also aware of the existence of various physical fields: gravitational, electric, magnetic. The presence of a field is indicated by its action on a particle immersed therein; to some extent we are led to believe that we understand the nature of a field when we can predict its action on a particle, although this is only partly true. We call the force on a unit particle (i.e., a unit mass in the case of the gravitational field or a unit charge in the case of an electric field) the *intensity* of the field.

We know that, in space, the field intensity of a point source will obey an inverse square law. This follows from our mental picture of the field as something emanating from the source point almost like a kind of liquid. That is, we imagine the "substance" involved moving outward from the point source in such a way that its total "amount" is conserved — through any sphere of radius r (and therefore area $4\pi r^2$) we see the same amount pass. By symmetry, then, the intensity must fall off as $1/r^2$. Furthermore, involved in the field expression is a numerical factor that remains constant (this is the gravitational constant times the mass of the point source in the case of the gravitational field, or a constant times the charge of the point source in the case of the electrostatic field). In cases where a medium offers some "resistance" to permeation by the field, the corresponding numerical factor is not constant and, moreover, we can lose the inverse-square nature of the field (the dependence can become $1/r^{2+\varepsilon}$, where $\varepsilon > 0$). Although we shall not attempt to discuss the various other fields that can act between elementary particles, we take note of the following: all point-sourced fields known to physics display a nonlinear dependence on distance from the source.

When atoms are grouped to form a solid or liquid body, two types of mutual forces appear among them: attractive and repulsive. The laws for these differ and serve to define stable relative positions of the atoms or molecules. Of course, the real picture is greatly complicated by random thermal motions of the particles, as well as by motions of their even simpler constituent particles. As a first approximation, however, it is common to regard the atoms as stationary when considering the internal structure of materials. One can find pictures and explanations in textbooks of how atoms constitute crystalline structures. Only the simplest qualitative explanations are given for these structures, primarily for the case of bodies composed of larger molecules such as polymers. In all cases, the internal forces of interaction that define the structure of a material are nonlinear in nature. So the question arises: if all laws of interaction are nonlinear, then how can we use linear equations to describe problems of elasticity or electromagnetism? The answer is as follows. All "real" physical laws seem to be nonlinear, but in many cases these can still be approximated by linear relations to a good degree of accuracy. Here we make use of the fact that, in a small neighborhood of a point, the behavior of any function can be approximated using its first differential. The first differential is linear in the increment of the arguments of the original function; the smoother the original function, the larger the range over which the approximation holds. Of course, linearity is a good model only over some particular range of increments. For example, the range of strains over which steel behaves elastically (returning to its initial state upon removal of the load) is of order 10^{-3}. To engineers who can live with errors of several percent, the results of linear elasticity are often satisfactory over the working range of a material. But *dislocations* — microscopic defects in the crystalline structure of a material — can introduce irreversible effects into the process of deformation. For steels, these occur near the edge of applicability of elasticity theory. Special technologies allow engineers to produce small amounts of "perfect" materials (e.g., carbon threads) that possess great strength. For these, the limit of elasticity is much higher than usual, and experiments show that the elastic range for such materials cannot be covered solely by the linear theory. There are also materials for which the elastic model fails for any level of strains. We are fortunate that this is not the case for ordinary materials, and that the classical theory of elasticity can be applied in practice.

3.8 Dislocations, Plasticity, Creep, and Fatigue

We have seen why a large part of structural behavior is covered by the theory of elasticity — indeed, by the linear theory. We have stated that this is because, in real life, the deformations in the working area under load are so small that the atomic structure of the material is not changed. (Our present discussion will apply to materials having relatively simple atomic lattice structures, like those of metals. For other materials, the arguments are more complex and normally only qualitative in nature.) Atoms of the lattice are shifted with respect to one another, but the general lattice structure is practically unaltered. If the material structure were perfect, the possible distortions of the lattice could be so large that only the nonlinear theory

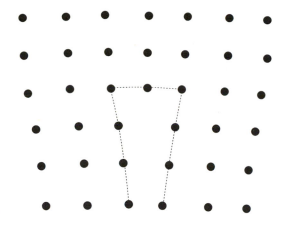

Figure 3.6 Linear dislocation in a crystal, cross sectional view. The dashed lines
indicate the border of the half-plane of atoms that was removed to form
the dislocation.

could describe them. Lattice defects do not permit higher strains to occur in
the nonlinear regime, and bring us to consider plasticity, creep, fatigue, and
related effects. These are caused mainly by elementary lattice imperfections:
the abovementioned dislocations. We shall describe the two simplest types
for a cubic-type lattice. We start by supposing that, for some reason, all the
atoms belonging to some half-plane within the lattice have been removed.
This would leave a gap where the missing atoms used to be. The edge of
the half-plane is a line along which a dislocation will be said to occur. For
definiteness, we shall imagine this line to be oriented vertically.

- Now, imagine the gap simply closing horizontally by virtue of attraction
  between the atoms on either side. As a result, the lattice returns to its
  original cubic form everywhere except near the vertical dislocation line.
  The zone of lattice deformation is an *edge dislocation*. See Figure 3.6.
- Alternatively, start with the gap open again. Two half-planes of atoms
  border the gap. Imagine that one of these planes, along with all of the
  atoms behind it, incurs a vertical displacement by a distance equal to
  one lattice spacing; all the atoms on the other side of the gap remain
  stationary. So all atoms on one side have been shifted longitudinally,
  along the dislocation line, by one lattice position. Finally, imagine the
  gap closing horizontally as before. The lattice returns to its original
  cubic form except near the dislocation line, and the resulting irregularity
  near that line is a *screw dislocation*. If the body is finite, then on
  the top and bottom surfaces we see a step-type discontinuity along the
  horizontal half-line where the relative shift occurred between the two
  sections. See Figure 3.7.

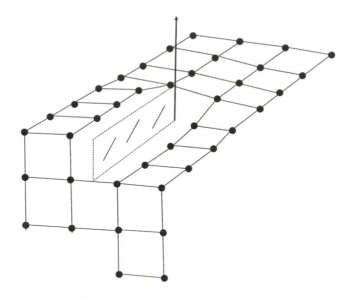

Figure 3.7 Representation of the atoms near a screw dislocation. Hatched plane
         indicates vertical step discontinuity in atomic locations at the crystal
         surface.

Although we have assumed straight-line dislocations above, in reality they
can be curvilinear. The two types can also combine into more complicated
kinds of dislocation. However, the fact of real importance is that, along a
dislocation, only a relatively low level of strain is required to cause some
atoms in the lattice to jump to other stable positions. Upon removal of the
strain, the lattice may or may not return to its prior configuration. When
few dislocations are present, we do not see the effects of this because they
are small. But the presence of a great many dislocations can cause us to see
a macroscopic distortion of the material after we remove the load. It is this
phenomenon that we call *plasticity.*

     The same effect can lead to *creep*. Here, the effects of changing dislocations
migrate slowly through the material under load. The resulting slow change
of shape that develops over time makes the material behave something like
a liquid, but with a small change in the strains.

     The greater the number of dislocations, the more plastic a material be-
comes. At first glance, it might seem that plasticity would be harmful and
that engineers should design structures in such a way that they function in
their elastic regimes only. But plasticity is useful in many circumstances.
This is the case, for example, when a structure is placed under a load that
does not change over time. Portions of the structure can enter the plastic
state and deform in ways that compensate for imperfections in the design.
After that, the structure can support the load. An example of this occurs with nuts that have been tightly screwed onto bolts having imperfect

threads: here, parts of the threads are deformed plastically but maintain the load perfectly, until the owner needs to unscrew the nut. Often the thread is fully damaged in the latter process. Because many structures are meant to support a permanent load, some parts of these can well be in a plastic state.

Another useful effect resulting from the accumulation of dislocations is referred to as *hardening*. This happens with plastic materials because of a redistribution of dislocations in such a way that the integral structure of the material is "improved" by the motion of the dislocations under load. In this way, we obtain a material that has better strength properties. The effect is widely used in engineering practice.

However, plasticity — reflecting as it does extremely complex changes in the internal structure of a material — is not described as neatly as elasticity. There are many approximate theories of plasticity, with some practically replicating the theory of liquids, and others resembling nonlinear elasticity (at least for some part of the process of plastic deformation). Plastic deformations of the same material under various loads should be described by different models of plasticity, so engineers need some practical experience in order to select the appropriate version for doing strain calculations.

The loading diagrams for many materials, such as steel, display some plasticity regime. But other materials, such as diamond, display no such regime. In the latter case, the elastic regime ends when the material is damaged under load.

One of the most important aspects of the strength of materials — the problem of the real strength of designed structures — is presented vaguely in textbooks. This is because the process of material damage is initiated on the internal structural level and develops from there — in large part, we see only the final results. Continuum mechanics uses integral characteristics of the processes inside the material, and we lack appropriate models that can predict precisely what happens under various internal conditions. The stress level, of course, often reflects the appearance of additional dislocations or other imperfections in the material structure. However, as with any integral characteristic, it cannot tell us precisely whether a structure will survive intact. In textbooks on the strength of materials, the reader will find several criteria that can be employed to test whether a material is being used in its safe operating regime. The multiplicity of such criteria is explained by the differences in materials themselves, and by the fact that different circumstances exist for loading a structure.

The modern viewpoint assigns responsibility for damage to a solid body to the presence of a large number of imperfections — of dislocations, in particular — in the internal structure of the material.

The mechanism of rupture of a sample under load can be described as the final result of the accumulation of singularities within the crystal lattice. With relatively few dislocations present, a material behaves elastically and offers resistance to loads. An increase in load brings an increase in strain; the greater the strain, the more dislocations that appear. However, at a certain stage, because of the effect of stress concentration, dislocations concentrate near singular points of boundaries, near points of inhomogeneity of the material, and near holes or places where the body is clamped. Along welded lines, in particular, the structure of a material is compromised by high changes in temperature. When the rate of formation of dislocations exceeds some threshold level, we will see microscopic cracks begin to appear, and a further increase in that rate will produce cracks developing in various ways, depending on the material. Some cracks grow slowly over a period of years, as we see in many old buildings; others develop catastrophically: a crack in a high-pressure pipe can grow at the speed of sound, opening up many meters of the pipe nearly all at once. Sometimes the presence of cracks means we must remove a structure from service, sometimes it does not. The evaluation of a structure's lifetime is somewhat of an art form, in which it is necessary to have engineering intuition, experience, and a more detailed knowledge of the actual stresses than is available from textbooks.

Although the strength of materials has attempted to cope with the issue of cracks for hundreds of years, it contains no absolute theory or tool that we can use to predict exactly how a crack will propagate or how long a structure will survive for any material. Most troublesome is that we have sparse experimental data for such dangerous structures as atomic reactors, because the material behavior under radiation and temperature extremes will differ from that under ordinary conditions. Even the figures of safety put forth by the designers of such facilities remain in question. Nobody can guarantee that a structure is really safe. Our meager understanding of the micromechanisms of crack development and propagation in metals really offers no absolute macrocriteria for the safe operation of real-life structures; all available criteria are probabilistic in nature. Existing criteria with which engineers determine whether, say, parts of an aircraft can continue to serve with imperfections found are quite specific and take into account the type of material, the technology of their development, and experimental data collected over many years.

The theory of cracks is studied in both mechanics and physics, but by differing approaches. Whereas mechanicists try to use the macro approach, physicists try for a statistical description. Unfortunately, the mechanisms of crack development are so complex that no universal theory exists to describe them: in large part, qualitative descriptions are available on how typical cracks arise and develop in typical materials. Material rupture near critical loading obeys only statistical laws: under the same load, two practically identical samples can survive for very different time lengths. Sometimes a negligible change in the technology of production (negligible even from the viewpoint of an experienced engineer) can bring a crucial change in the lifetime of a loaded structure.

We are now aware of how stress tends to concentrate near irregularities. At these points, cracks have much higher chances to arise and develop. The surface of any heavily loaded object is dangerous from this viewpoint. First of all, any mechanical processing of a material will cause surface irregularities. Cutting tools leave traces, and high-temperature processes leave changes behind as well. Careful polishing is sometimes done for aesthetic reasons, but is often intended to remove imperfections and thereby increase the strength of a structure. We know, for example, that a small scratch on a glass can cause it to shatter. A similar thing can happen to a steel structure, but the process is slower and therefore not often attributed to small surface imperfections. Soon, we shall discuss fatigue in materials; this makes scratches, small holes, and other shape imperfections much more dangerous in structures subjected to cyclic loading. Even a one-millimeter scratch made in a truck beam, using a small file, can cause the beam to fall apart after thousands of loading cycles. On bad roads, this does not take too long.

We know a lot, but still comparatively little, about how material damage occurs. The problem is quite complex. We lack extensive experimental data for actual structures because the scale models often used to collect data are not really representative. There was one example of a real-life experiment, however, that provided a lot of material for science: the "Liberty fleet." Several thousand ships bearing this designation were produced circa World War II for one-time transport of cargo across the Atlantic Ocean. Despite this limited intented use, the Liberty ships kept going for years. But then they suddenly began to sink, many being broken in half. On one of these ships, there was a chef[3] who found a crack that developed through his kitchen and then through his desk. The chef began to mark the end of the crack with dates and times, and followed its rather slow progress. He showed his results to the captain, and the story of these ships was so well known by that time that the captain tried to halt the crack by various methods used in similar circumstances (the drilling of perfectly round holes in the path of propagation, the strengthening of the desk, and so on). But the crack continued to propagate, and the captain and his crew were happy when they reached port before the ship split apart. Engineers studied the chef's markings, and the information was used to prevent similar cracks in future designs.

The abundance of mechanisms by which structures can incur damage is demonstrated by the fatigue of materials under cyclic loading. If a structure is loaded statically by some forces, it can survive for many years. However, if only half of such a load is applied cyclically for a thousand cycles, then it can destroy the structure. This happens to bridges. The growth of fatigue in a material is also explained by an increase in the number of dislocations. The oscillation of a load implies the appearance of dislocations in the crystal lattice at new places; these move, and so a new cycle of loading brings a new distribution of stress inside the material. Thus, new dislocations arise. The reader has become aware of this effect while trying to break a wire by repeated bending. Fatigue is dangerous in structures such as vehicles, bridges, and so on, and regular testing for cracks and other signs is often necessary.

---

[3]History seems to have forgotten his name, but this person contributed more to the theory of cracks than many professors working in the same area.

3.9 Heat Transfer

So far, we have omitted thermal effects from our mechanical models. Indeed, it is often reasonable to partition temperature and deformation effects into separate classes of problems. But in other cases this is impossible because a temperature change can affect the strain state significantly. Let us discuss the problem of temperature as it was done historically, beginning with phenomena that are independent of deformation processes.

A change in temperature of a body, if kept to a certain range, does not necessarily imply significant changes in other properties. So, temperature changes were at one time attributed to the presence of a fluidlike agent. The nature of this agent was never fully clarified, but the viewpoint was not too different from the modern one regarding electromagnetic or gravitational fields. The "fluid theory" of heat was abandoned when further developments in physics showed temperature to be a manifestation of the motions of the atoms and molecules out of which substances are composed. However, the picture of heat as something that flows like a fluid continues to provide the best available working model for problem-solving purposes. There is nothing wrong with making use of appropriate analogies. Indeed, the similarities in form possessed by the equations governing differing phenomena often yield a sound basis for the use of analogy. The two-dimensional Laplace equation holds in hydrodynamics and electromagnetism in addition to describing an elastic membrane, and therefore our experience with the latter system can help us understand what happens in other fields.

In fact, analogy is one of the most powerful ways to move forward in a new area of study. Here we want to point out the falsity of many rumors regarding how certain parts of science have been developed. Stories abound describing how equations came to their authors like flashes of lightning from the heavens. There is, of course, the famous story about Newton and the falling apple. But the complete history of the gravitational law is by no means that simple. Newton published the final result only after twenty-seven years of work on the subject. Moreover, his first version of the law was absolutely incorrect. This was pointed out to him in an extremely polite manner by Hooke, who, as we noted earlier, could be called the "Mozart of science." Hooke worked easily in any scientific area and produced new ideas in a way that no scientist has ever been able to duplicate. In his note to Newton, Hooke described the things that should be taken into account when deriving a law of gravity. This was done in such detail that Hooke could be considered as at least a coauthor of the law. It is possible that the ease with which Hooke dealt with complex problems introduced some tension into his relations with Newton. The final, and unfortunate, outcome is that today we know little about Hooke or his life. It is a sad commentary on human politics that someone could make such fundamental contributions in so many fields — from the basic law of linear elasticity to the cell structure of living organisms — and have even his final resting place veiled from succeeding generations. Turning to another example, Maxwell's equations of electromagnetism arose out of an analogy with hydrodynamics. This becomes evident to anyone who takes the time to read Maxwell's original manuscripts.[4] Only one idea — that of "displacement current" — was brought into the equations rather artificially in order to remove certain inconsistencies from the theory as a whole. It took Maxwell a long time to introduce this idea into the fundamental equations that were to eventually bear his name.

So, we hope we have convinced the reader that it is valid to have in mind fluid-flow pictures when considering heat transfer problems. Of course, the ultimate justification is that the results are practical and confirmed by actual experimentation.

Let us begin by considering how to measure an "amount" of heat Q. Our fluid analogy suggests that the increase in heat carried by a body should be proportional to the mass m of the body and the change ΔT in its temperature. That is, we should be able to write $\Delta Q = cm\Delta T$, where c is the coefficient of heat capacity. We normally write $m = \rho V$, where V and ρ are the volume and density of the body, respectively; so we have

$$Q = c\rho V \Delta T. \tag{3.11}$$

Thus, the increment of heat is linearly related to the temperature increment.

Further experimentation showed that Q can be considered as a type of energy of the body. Indeed, it was found that the heat carried by a gas could be transformed into work done by forces, and vice versa. This was a great achievement along the path to revealing one of the greatest laws of physics: the law of energy conservation.

We now begin to consider how the equation of heat transfer is derived. Fourier[5] discovered a law that explains that the amount of heat passing axially through a cylinder is proportional to the difference in temperatures between the endcaps. Limiting considerations brought him to the idea that the amount of heat passing uniformly through some flat area S with normal directed along the x-axis is proportional to the spatial rate of change of the temperature:

$$q = -k\frac{\partial T}{\partial x},$$

where the negative sign reflects the fact that heat passes from regions of higher temperature to regions of lower temperature. During the time interval $[t, t + \Delta t]$, there is a transfer of heat given by

$$Q_1 = -kS \int_t^{t+\Delta t} \frac{\partial T(x, \tau)}{\partial x}\, d\tau. \tag{3.12}$$

Regarding heat as a kind of gas that cannot appear or disappear (except

---

[4] James Clerk Maxwell (1831–1879).
[5] Jean Baptiste Joseph Fourier (1768–1830).

at the locations of external sources whose densities we will describe using a function F), we can derive a heat transfer equation for a slab characterized by its cross section S. We suppose that over each cross section the temperature distribution T is constant, and so T can be considered as a function $T = T(x,t)$, where x is the local longitudinal coordinate. We consider a small slice of the slab between the points x_0 and $x_0 + \Delta x$ during the time interval $[t, t + \Delta t]$, and use the conservation of heat in the volume. By (3.12), the net input of heat from outside this part of the slab is given by

$$-kS \int_t^{t+\Delta t} \frac{\partial T(x_0, \tau)}{\partial x} \, d\tau + kS \int_t^{t+\Delta t} \frac{\partial T(x_0 + \Delta x, \tau)}{\partial x} \, d\tau$$

$$+ S \int_t^{t+\Delta t} \int_{x_0}^{x_0 + \Delta x} F(x, \tau) \, dx \, d\tau.$$

The heat accumulated in this part of the slab during $[t, t+\Delta t]$ is, by equation (3.11),

$$S \int_{x_0}^{x_0+\Delta x} c\rho T(x, t + \Delta t) \, dx - S \int_{x_0}^{x_0+\Delta x} c\rho T(x, t) \, dx.$$

So, the heat conservation assumption gives us

$$k \int_t^{t+\Delta t} \left[\frac{\partial T(x_0 + \Delta x, \tau)}{\partial x} - \frac{\partial T(x_0, \tau)}{\partial x} \right] d\tau + \int_t^{t+\Delta t} \int_{x_0}^{x_0+\Delta x} F(x, \tau) \, dx \, d\tau$$

$$= \int_{x_0}^{x_0+\Delta x} c\rho \left[T(x, t + \Delta t) - T(x, t) \right] dx.$$

This is the equation of heat transfer for the slab. A differential form can easily be derived through an application of the mean value theorem for integrals. We have

$$k \left[\frac{\partial T(x_0 + \Delta x, t + \theta_1)}{\partial x} - \frac{\partial T(x_0, t + \theta_1)}{\partial x} \right] \Delta t + F(x_0 + \xi_1, t + \theta_2) \Delta x \Delta t$$

$$= c\rho \left[T(\xi_2, t + \Delta t) - T(\xi_2, t) \right] \Delta x,$$

where $0 \le \theta_1, \theta_2 \le \Delta t$, and $0 \le \xi_1, \xi_2 \le \Delta x$. Dividing this through by $\Delta x \Delta t$ and passing to the limit as $\Delta x \to 0$ and $\Delta t \to 0$, we obtain

$$\frac{\partial}{\partial x} \left(k \frac{\partial T}{\partial x} \right) + F = c\rho \frac{\partial T}{\partial t}.$$

This result for a slab can be extended to describe heat transfer in space when k is constant:

$$\frac{\partial T}{\partial t} = a^2 \Delta T + F. \tag{3.13}$$

So, we now have the governing equation for the spatial variation of temperature. As with any differential equation, we need to impose additional conditions on T in order to get a unique solution. A well-defined boundary value problem arises if we appoint a temperature T at the instant $t = 0$ and supplement this with a boundary condition for $t > 0$. The latter condition must describe how the temperature of the boundary changes over time or how heat passes through the boundary. The resulting problem has a unique solution (within some class of functions) that depends continuously on changes in the initial or boundary conditions and the external forces.

The presentation in almost any textbook will obscure the fact that, during the development of a theory, attempts were made to consider and solve other types of problems: the reader sees only the results of the more "successful" efforts. It is interesting to note that in the history of heat transfer, a certain class of problems was considered unsuccessfully and then reconsidered much later when more powerful tools had become available. These were the *inverse problems*, in which the final temperature distribution in a body was given and the entire history of the distribution was sought. Such problems turn out to be *ill posed*: a small disturbance in the final temperature distribution can give rise to huge disturbances in the solution for the temperature history. In recent decades, inverse problems have attracted the attention of persons working in the theory of differential equations. One result has been a sequence of advances in the area of spatial mapping known as tomography.

It is worth understanding why the heat transfer equation is so sensitive to the direction in which time progresses (since Newton's dynamical laws display no such sensitivity). Formally, the replacement of t by $-t$ in equation (3.13) causes a sign change in the term on the right; this means that ordinary solutions, which for positive time vary as $e^{-\lambda t}$, will, after the transformation, vary as $e^{\lambda t}$. In the Fourier method of solution of the heat transfer equation, we make use of numbers λ_k that tend to ∞ as $k \to \infty$, and in this way we get the instability effect mentioned above.

This provides a formal explanation of why the inverse problem of heat transfer is ill-posed.

3.10 Thermoelasticity

The heat transfer equation we have derived works nicely when applied to a body whose shape is unaffected by temperature changes. However, in the physics of gases, it was shown that a change in temperature relates to changes in pressure and volume. For solid bodies, temperature effects on shape are less pronounced, but often cannot be ignored. First, we would like to understand how stresses and strains relate to temperature — in other words, how to incorporate thermal effects into Hooke's law. Elementary physics supports our everyday observation that a body will increase its dimensions in response to an applied temperature increment ΔT. In the linear approximation, we have

$$\Delta L = \alpha L \Delta T,$$

where L is the linear dimension under consideration. For steel, the coefficient of thermal expansion is about 10^{-5} in SI units. The factor α is positive for many materials but not all. For example, if we stretch a resin band and then heat it, we will see it shorten. The explanation for this "unusual" behavior is as follows. Resin molecules are long and tend to get tangled up with their neighbors. The resulting "balls" are elongated upon stretching of the band; upon heating, they offer more resistance to elongation. Such qualitative explanations are, of course, only tentative. As a similar example, we have no elementary explanation for the fact that the coefficient of thermal expansion for water under normal atmospheric pressure is negative up to $4°C$ and positive at higher temperatures.

Let us return to the above relation. We see that $\alpha \Delta T$ defines the change in strain $\varepsilon = \Delta L / L$. For the linear version of Hooke's law, written in one-dimensional form for a material under constant temperature, we have $\sigma = E\varepsilon$; after a temperature change ΔT, we should exclude the strain $\alpha \Delta T$ that does not imply a change in stress:

$$\sigma = E(\varepsilon - \alpha \Delta T).$$

For steel, a change $\Delta T = 100°K$ gives a thermal strain of about 10^{-3}. Such a strain level will put a simple steel near the boundary of its range of applicability if the strain is due to loading. So, thermal expansion can require significant attention during the design of ordinary structures that are to be exposed to outdoor temperature excursions of $50°C$ or more.

The extension of this law to three dimensions is aided by the experimental fact that for a homogeneous material the mixed components of the stress tensor do not depend on temperature changes. Because homogeneity also implies a directional independence, we can write the constitutive law of thermoelasticity in Cartesian coordinates as follows:

$$\sigma_{ij} = [\lambda\theta - (2\mu + 3\lambda)\alpha\Delta T]\delta_{ij} + 2\mu\varepsilon_{ij}. \tag{3.14}$$

Substituting this into the equations of equilibrium in terms of the components of the stress tensor, we will get the equations of equilibrium in displacements and temperature. If we know the temperature field in a body, then the terms with ΔT are known, and we can solve the corresponding boundary value problem as a usual problem of elasticity with given "external forces" determined by the temperature change. There is a mutual interaction between the stress field and the temperature distribution in a solid body, but we can often neglect this and solve the simplified problem. First, we can solve for the temperature distribution, then for the strains and stresses. Next, we will discuss some practical applications of thermoelasticity.

3.11 Thermal Expansion

The effects of thermal expansion can be either harmful or helpful. Internal stresses can arise in a thick glass vase through nonuniformities in both solidification and in the temperature distribution of the molten material before solidification. Items composed of materials having differing coefficients of thermal expansion may be susceptible to damage when exposed to temperature variations. But the same effect is used to advantage in thermally controlled switches: here, thin strips of two different materials are bonded together so that the composite strip will bend in response to temperature changes.

When designing objects to be made from different materials, engineers normally try to choose materials having thermal expansion coefficients that are not too far apart. One such pair of materials was discovered by chance when someone installed an iron frame into a concrete flower vase. The structure turned out to be so successful that now almost any concrete detail in a building will contain iron wires as structural reinforcements. It is interesting that the thermal expansion effect is used even more intensively for many structures composed of concrete and steel. Concrete is a material whose stress–strain diagram is nonsymmetric with respect to tension and compression: it can withstand significant compression but is damaged much more easily under tension (and shear). Bending resistance is especially important when concrete plates are used to form, for example, the floors of balconies. The idea is to prestress the concrete plate in such a way that during bending the part that would normally be stretched is merely relieved of its stress. This is accomplished by first heating a steel frame and then pouring the concrete around it: when the frame cools, it contracts, and thereby compresses the nearby concrete.

Thermal expansion allows us to fit a carefully made metallic pipe inside another pipe whose inner radius is exactly the same as the outer radius of the first pipe: we simply heat the external pipe until it expands sufficiently. The same technique, applied when the inner radius of the external pipe is a bit smaller, allows us to firmly clamp two pipes together. This trick brought crucial changes to the design of large guns. The oldest gun barrels had to have extremely thick walls in order to withstand the pressure of exploding gunpowder. With the advent of steel, the wall thickness could be reduced but remained significant. However, analysis of the strain distribution in a thick barrel under internal pressure reveals that the greatest strain intensities occur near the inner surface; at some distance from this surface, the material is practically unloaded. The strain on the inner particles was eventually reduced through the introduction of a radial prestress: an initially compressed state could absorb enough of the explosion pressure to preserve the integrity of a thinner barrel. The first implementation of this idea involved the use of hoops (like those found on wine barrels), but a later construction method relied on the tight-overlap layering of separate pipes, as discussed above. Into a strongly heated outer pipe was inserted a cool inner pipe; after cooling, the outer pipe served to compress the inner pipe as desired. The thickness of gun barrels was significantly reduced in this way, and the technique continues to find extensive application in industry.

3.12 A Few Words on the History of Thermodynamics

The isolation of the heat transfer problem for solids makes sense in those situations where extremely accurate results are not needed, but there are problems for which the strain and temperature effects cannot be kept separate. The reader is aware that repeated bending of a wire in the same spot can increase the temperature of that spot significantly. The isolated models cannot display this effect in principle. A very precise measuring instrument may have to be designed with the mutual interaction of strain and temperature in mind. This interaction can be described using the tools of *thermodynamics*, a science that arose out of the study of the gas laws. Thermodynamic effects of various natures can be inserted into a model as necessary. At this point, we consider thermodynamics to be a field residing within continuum mechanics.

The relation of heat problems to mechanical problems was suggested by experiments showing that the work of forces can be transformed into heat, and vice versa. So, the quantity of heat can be considered as a type of energy possessed by a body. This provided the background for the introduction of heat problems into mechanics. In the history of thermodynamics, two names stand at the pinnacle: Sady Carnot (1796–1832), who formulated the basic statements of thermodynamics, and J. W. Gibbs (1839–1903).

Sady Carnot[6] studied the efficiency of a steam engine. This was a practical problem motivated by the crude machines of his time: it had become necessary to seek the best regimes and modes of operation. Carnot not only found the best solutions; he also managed to formulate a basis for thermodynamics in such a way that, for over a century, all the best textbooks on the subject have merely found new ways to rephrase his statements. As with many pioneers, Carnot wrote slightly over the heads of his contemporaries. He formulated what we now call the second law of thermodynamics (the first law being that of energy conservation). The second law was stated in around twenty forms that Carnot regarded as equivalent. Many were subsequently attributed to other persons. Carnot considered a cyclic process performed by an engine. One of the cycles would result in heat being transferred by steam from an "infinite" heat source, or *capacitor*, — at a higher temperature to another capacitor at a lower temperature. During the transfer process, the steam would perform some work, the infinite nature of the heat capacitors being necessary for their temperatures to remain constant during the cycle. This is called the "direct cycle" of the steam engine. Conversely, the transfer

---

[6]Sady Carnot was a son of Lazare Canot. The elder Carnot was a mathematician, mechanicist, and engineer; he was also Napoleon's minister of war for a period of five months. Lazare Carnot educated his son and had great influence on him.

of heat from a cooler capacitor to a hotter one would require the expenditure of work; this is the result of the "inverse cycle." Suppose we were to take two identical engines and connect them in such a way that the first engine does work, transferring heat from a hotter source to a cooler one (direct cycle), while the second engine functions in the inverse regime and uses the work done by the first engine to transfer heat from the cooler source back to the hotter one. Each engine is said to be *perfectly reversible* if the pair causes no net change of heat content in either heat source over the course of a cycle of operation. Carnot thought this was impossible, and this is one of his formulations of the second law of thermodynamics. He regarded it as impossible for any engine to perform actions resulting solely in the transfer of heat from a cooler source to a hotter source. Any engine that supposedly violates this law could be termed a "perpetual motion machine of the second kind." The idea of creating such a machine (as well as a "first-kind" machine that could violate the law of energy conservation) has captured the imagination of many unwary inventors over the years. Let us note that, unlike the first-kind machine, the second-kind machine is forbidden only on statistical grounds — for example, when dealing with large volumes of working gas or with many cycles of operation. In any volume of a gas, there are atoms with various amounts of energy, and it is conceivable that the atoms with higher energies could be selected by chance and moved by an engine to a hotter room. But such a violation of the second law of thermodynamics is extremely improbable. Carnot was therefore safe in applying the mathematical "axiom game" in an area of physics: he took a rule that seemed correct, supposed it to hold in an absolute sense, and considered the consequences. One consequence was that the efficiency of a perfectly reversible engine is the highest of all possible engines. In this way, the notion of absolute temperature was introduced along with a mathematical formulation of the second law of thermodynamics. In particular, the notion of entropy as a characteristic property of materials appeared.

Later workers clarified and extended the contributions of Sady Carnot. This gave a push to both physics and chemistry. Gibbs considered gases as great collections of atoms moving chaotically and, using the tools of statistics, he clarified the meanings of the laws and functions of thermodynamics. But statistical physics, while dealing successfully with gases, gives only qualitative results for solids, and this is still an open problem. Besides many chemical applications of thermodynamics, Gibbs also considered the question of stability of thermo-processes, and the arguments he originated persisted for decades. The first applications of thermodynamics to the mechanics of solids were also given by Gibbs in his famous work on the equilibria of heterogeneous substances.

The second law of thermodynamics is formulated for irreversible processes as well, but in this case its mathematical statement is an inequality that is hard to use in calculations. A theory of locally reversible thermodynamics has been applied to such processes. The process is then globally irreversible, but is locally described by equations instead of inequalities.

For irreversible processes, the situation is analogous to the laws of plasticity. The effects are complex and cannot be described by one simple law. Heat is dissipated and converted into other forms of energy, so for each irreversible process it is necessary to use special approximations for the dissipation or conversion that account for peculiarities in the process.

3.13 Thermodynamics of an Ideal Gas

Elementary physics treats the thermodynamics of a gas using just a few parameters: the pressure p, the volume V, and the absolute temperature T. The state of the gas is assumed to be homogeneous so that it can be described by integral parameters characterizing the gas as a whole. At any moment, the gas is in equilibrium, hence this theory could be termed "thermostatics" instead of thermodynamics. The parameters of an ideal gas are related by a constitutive equation such as

$$pV = \frac{m}{\mu}RT, \tag{3.15}$$

where m is the mass of the gas, μ is the molecular mass of the gas, and R is the gas constant.

We see that only two of the three descriptive parameters are independent. In fact, we could employ any pair of parameters that could fulfill the role of coordinates in classical mechanics. The number of independent parameters (here, two) is analogous to the number of degrees of freedom that characterizes a system in classical mechanics.

In elementary physics, we are not concerned with the mechanisms by which the parameter changes are sequenced. We usually consider particular processes in which one of the parameters remains constant. When $T = $ constant, the process is called *isothermal*; when $p = $ constant, it is *isobaric*; when $V = $ constant, it is *isohoric*. The isothermal process is usually visualized as occurring when a gas is in contact with an extremely large heat reservoir at constant temperature. Heat is transferred from the reservoir to the gas, or vice versa, in such a way that the reservoir temperature does not change. How this is accomplished is outside the scope of such a description, and the heat transfer equation is not applicable. This observation holds for any of the processes in a gas described by equations of the type (3.15). The state of a gas during any of the elementary processes mentioned above is fully determined by a single parameter.

There are other types of elementary gas processes. So far, we have not considered heat explicitly. To do so we must write out the balance equation for the change of heat δQ, the internal energy δE, and the work δW done by the external forces acting on the gas, for two infinitesimally close equilibrium states of the same mass of gas. This equation is what we call the law of conservation of energy for the gas:

$$\delta E = \delta W + \delta Q. \tag{3.16}$$

It can be said that the change in internal energy of the gas is due to the sum of the work done by external forces deforming the system and the addition of heat into the system. Because we consider equilibrium states of the system, the change in kinetic energy is neglected. In elementary physics, the term δW for a gas is $p\, dV$; for spatial objects, it should be formulated in accordance with the results of continuum mechanics. The energy of a system is a function of the system's internal parameters, and so it is uniquely defined by the values of those parameters. However, we never know the absolute value of the energy function; we deal only with its increments. It can be formulated pointwise for a distributed system, but we watch, in large part, the change in the integral value of the energy due to the change in those parameters we have appointed as the principal ones. Because E is a function of the internal parameters, the main part of the infinitesimally small change in energy (over infinitesimally small changes in the internal parameters) is its first differential dE. Because the equation is written for infinitesimally close states of the gas due to an infinitesimal change of the system parameters, in (3.16) we can change δE to dE. In contrast, the expression δW cannot be represented as a complete differential of some function; otherwise, in any cyclic process we would get zero work done by the system. This, of course, would contradict our experience. The same holds for δQ: it cannot be a complete differential of a function of the internal parameters. However, for reversible processes for a δQ, we can define a new function of the internal parameters.

For an ideal gas obeying the law (3.15), it turns out that if we calculate the quantity

$$\int_{(A_1)}^{(A_2)} \frac{\delta Q}{T}$$

along any "path" from state (A_1) to state (A_2), we find that it does not depend on how heat was delivered to the gas. Instead, it depends only on the initial and the final states. This happens for any path. From calculus, we know this means that the elementary quantity $\delta Q/T$ is the differential of some function of the internal parameters of the system (gas). This function is called the *entropy* and is denoted by S. The change of the entropy function is zero during a cycle, when a system returns to an initial state. Thus S is a function of the internal parameters. Carnot has told us that all reversible engines have the same efficiency. It follows that the idea of entropy is universal, that it can be introduced for any reversible "engine" regardless of the working body involved.

We can define only the change in entropy of a body. In quantum mechanics there is a so-called third law of thermodynamics, which states that at zero absolute temperature, the entropy is zero. In this way, we can talk about absolute figures for the entropy.

We know that two internal parameters, T and V, for example, fully define the state of an ideal gas. Knowledge of the entropy as a function of internal parameters allows us to introduce other pairs of independent parameters, V and S, for example, uniquely defining the state of the gas.

Thermodynamics uses primitive tools, but for the description of processes, we can use various independent variables and notions that are not introduced with absolute strictness — like the notion of isolated system, among others. This is why thermodynamics is often thought of as a complicated science (students often judge a complicated problem to be "simple" if it is amenable to only one approach; conversely, they will judge a simple problem to be "complicated" if many possible approaches present themselves).

Finally, let us note that in addition to the main elementary processes mentioned above, there is an important type of process called *adiabatic*. Such a process occurs when a thermodynamic system is isolated from external heat sources. For an ideal gas undergoing an adiabatic process, the conservation law with $\delta Q = 0$ implies an additional relation (a differential equation), from which it follows that a single internal parameter, V, for example, uniquely determines the state of the gas. If we consider V and T as the main descriptive parameters, a change in V always uniquely defines some change in T in an adiabatic process. This kind of elementary process is important in practice.

For irreversible engines, entropy is introduced as well, but rather than discussing this, we will proceed to the thermodynamics of an elastic rod. The rod can be considered analogous to a gas, being a working element of a reversible engine.

3.14 Thermodynamics of a Linearly Elastic Rod

For an elastic rod, we take the constitutive equation

$$\sigma = E[\varepsilon - \alpha(T - T_0)], \qquad (3.17)$$

where T_0 is an initial temperature. For simplicity, we consider a unit cube, so the stress σ in the cross section is numerically equal to the force that stretches the rod, and ε becomes the simple elongation of the rod when uniformly stretched. The most appropriate internal parameters with which to characterize the rod are ε and T.

The energy conservation equation (here we denote the energy by $\mathcal{E}$ because we reserve E for Young's modulus) $\delta\mathcal{E} = \delta W + \delta Q$ becomes

$$\delta\mathcal{E} = \sigma d\varepsilon + \delta Q, \qquad (3.18)$$

because the work of the stretching force σ over an additional elongation $d\varepsilon$

is $\sigma d\varepsilon$. We consider the processes of stretching and heating the rod when at any moment the rod is in equilibrium. The work done by the stretching force during a cycle (a cycle is finished when ε and T return to their initial values) is zero. Taken as a requirement, this could serve to define the term "elastic rod." Meanwhile, it means that we now deal with a reversible-engine-type element; we can therefore introduce, as for the ideal gas, the entropy function S by the equation

$$\delta Q = T\, dS.$$

This is the formulation of the second law of thermodynamics for the rod. Thus, the expression for the law energy conservation is

$$\delta \mathcal{E} = \sigma d\varepsilon + T\, dS. \tag{3.19}$$

As for a gas, we can characterize the rod by two internal parameters: ε and S. In (3.19), the expression for $\delta\mathcal{E}$ in terms of ε and S becomes a complete differential; hence, comparing (3.19) with the definition for the first differential

$$d\mathcal{E} = \frac{\partial \mathcal{E}}{\partial \varepsilon} d\varepsilon + \frac{\partial \mathcal{E}}{\partial S}\, dS,$$

we get the relations

$$\sigma = \frac{\partial \mathcal{E}}{\partial \varepsilon}, \qquad T = \frac{\partial \mathcal{E}}{\partial S}.$$

So $\mathcal{E}$ serves as a potential function for the pair of parameters σ and T.

We can treat other pairs of parameters as independent. Corresponding to the common pair ε, T, we can introduce another function called the *free energy* $F(\varepsilon, T) = \mathcal{E} - TS$. In terms of F, equation (3.19) becomes

$$dF = \sigma d\varepsilon - S dT.$$

From this, we conclude that

$$\sigma = \frac{\partial F}{\partial \varepsilon}, \qquad S = -\frac{\partial F}{\partial T}. \tag{3.20}$$

In a similar way, we can introduce potentials for other pairs of independent variables. For the pair σ, S, we use the *enthalpy* given by

$$\mathcal{I}(\sigma, S) = \mathcal{E} + \sigma\varepsilon,$$

and the *free enthalpy* given by

$$\mathcal{I}^*(\sigma, T) = \mathcal{I} - TS.$$

In our earlier work with the theory of elasticity, we mostly considered cases where temperature dependence is not a concern. We essentially treated materials as though they remain at a constant temperature. In other words, only isothermal processes were considered; the elastic moduli we used were those that would be obtained at a constant temperature.

We wish to understand how a change in temperature can arise during deformation if there is no source of additional heat. So we are interested in adiabatic processes where $\delta Q = 0$, hence $S = $ constant. Let us assume that E, the isothermal modulus of elasticity, and α, the expansion coefficient, are constant and therefore independent of T. Next, we assume that when the strain $\varepsilon = 0$, the internal energy is

$$\mathcal{E} = c_\varepsilon T, \tag{3.21}$$

with a constant coefficient of heat capacity c_ε.

Under these assumptions (which seem to be faithful approximations), we can find expressions for the thermodynamic functions. Let us begin with relations (3.20), supplemented with (3.17). Thus,

$$\frac{\partial F}{\partial \varepsilon} = E[\varepsilon - \alpha(T - T_0)].$$

We need to restore F as a function of ε and T. Integrating the last relation with respect to ε, we get

$$F = \frac{E\varepsilon^2}{2} - E\alpha\varepsilon(T - T_0) + \phi(T),$$

where $\phi(t)$ is an arbitrary function of T to be defined. Using the second relation from (3.20), we find the entropy

$$S = -\frac{\partial F}{\partial T} = E\alpha\varepsilon - \phi'(T)$$

and the internal energy

$$\mathcal{E} = F + TS = \frac{E\varepsilon^2}{2} + E\alpha\varepsilon(T - T_0) + \phi(T) - T\phi'(T). \tag{3.22}$$

When $\varepsilon = 0$, we get $\mathcal{E} = \phi(T) - T\phi'(T)$ and, so, by assumption (3.21), we get an ordinary differential equation for $\phi(T)$:

$$\phi(T) - T\phi'(T) = c_\varepsilon T.$$

This is easily solved, and its solution has an arbitrary constant. Requiring that $S = 0$ at $T = T_0$ and $\varepsilon = 0$, we obtain

$$S = E\alpha\varepsilon + c_\varepsilon \ln \frac{T}{T_0},$$

or, equivalently,

$$T = T_0 \exp \frac{S - E\alpha\varepsilon}{c_\varepsilon}. \qquad (3.23)$$

The equation for the energy $\mathcal{E}$ becomes

$$\mathcal{E} = F + TS = \frac{E\varepsilon^2}{2} + E\alpha\varepsilon(T - T_0) + c_\varepsilon T.$$

The relation (3.17) for σ is written for all cases. We wish to restrict it to adiabatic processes for which $S = $ constant. Because during the derivation we required $S = 0$ at $T = T_0$, $\varepsilon = 0$, we find that for the adiabatic deformation, $S = 0$ for all values of the parameters. This gives us

$$E\alpha\varepsilon + c_\varepsilon \ln \frac{T}{T_0} = 0,$$

from which, for the adiabatic deformation, we have

$$T = T_0 \exp \left(-\frac{E\alpha\varepsilon}{c_\varepsilon} \right).$$

Substituting this into the constitutive equation $\sigma = E[\varepsilon - \alpha(T - T_0)]$, we get

$$\sigma = E \left[\varepsilon - \alpha T_0 \left(\exp \left(-\frac{E\alpha\varepsilon}{c_\varepsilon} \right) - 1 \right) \right].$$

From this, we see that under adiabatic conditions, the relation between σ and ε is not linear — unlike the case for isothermal processes. But for small ε the influence of nonlinearity is small, and we can linearize this relation using a power series expansion of the exponential. Then we get

$$\sigma = E \left(1 + \frac{E\alpha^2 T_0}{c_\varepsilon} \right) \varepsilon. \qquad (3.24)$$

This is used to replace Hooke's law; thus, the coefficient $E(1 + (E\alpha^2 T_0)/c_\varepsilon)$ plays the role of Young's modulus for adiabatic processes. We recall that in an adiabatic process, the temperature changes according to equation (3.23). We can linearize this as well, remembering that $S = 0$, and get

$$T = T_0 \left(1 - \frac{E\alpha}{c_\varepsilon} \varepsilon \right). \qquad (3.25)$$

So the rod becomes cooler when stretched and warmer when compressed adiabatically.

Let us note that $\mathcal{E}$ in terms of the ε, S variables is

$$\mathcal{E} = \frac{E\varepsilon^2}{2} + E\alpha\varepsilon(T - T_0) + c_\varepsilon T_0 \exp \left(\frac{S - E\alpha\varepsilon}{c_\varepsilon} \right).$$

The expression for σ in the adiabatic process can be obtained by taking the partial derivative of $\mathcal{E}$ with respect to ε.

Let us discuss some consequences of this formula. For an adiabatic process involving small strains, we get a change of elastic modulus by the amount $(E\alpha^2 T_0)/c_\varepsilon$. For steel at room temperature, this introduces a change into the third significant figure of the stress value, and is often negligible. Of course, this depends on the situation: such an error made in weighing a 50-ton truck on an electronic scale could amount to several hundred kilograms, and could be expensive. Real processes are neither adiabatic nor isothermal, but rather fall somewhere between these processes. When loading occurs relatively quickly, the change of strains follows the adiabatic law. Then, over time, the strains relax to those defined by the isothermal law, because the temperature tends toward that of the environment. The relaxation process for the temperature looks similar to creep, and for steel is often taken as creep. For relatively small measurement elements, the time of "relaxation" of the temperature is relatively small, and we can read measurements comparatively quickly. For larger elements, the relaxation time can last for seconds or minutes, and this means that for precise weighing we need to wait, or use a program that recalculates instantaneous results to compensate for errors in readings.

3.15 Stability

We normally say that a system is *stable* (or exists in a stable state) if its behavior is steady and if slight disturbances in external parameters such as loads, shapes, and so on, lead to only slight changes in that behavior. This is close to what mathematicians refer to as the *well-posedness* of a problem: a small change in the external data should lead to a uniquely defined small change in the output (solution). When system behavior becomes irregular, we speak of a *loss of stability*. Many classical examples occur in connection with the resonance phenomenon. A certain bridge may support the weight of many large trucks rolling over it, but a much lighter group of soldiers marching in step with just the right rhythm will cause a major collapse. However, loss of stability may also be used to our advantage. The notion can be used to design control systems, and it can thereby serve to prevent catastrophes from occurring.

Irregularities take so many forms that it is difficult to precisely define what we mean by the term. In dealing with real objects, we must often judge that something is wrong by making observations that are hard to formalize. This is the case with loss of stability in mechanical objects. We noted earlier that the material presented in textbooks typically represents an extremely narrow selection of those results and ideas that have been judged as most successful. It would take many volumes to consider all stability-related problems that have arisen in the various branches of science. Two problems that have the same mathematical form may be treated using vastly different approaches by persons working in different areas. The best we can offer here is a quick sketch of the more famous stability problems that occur in mechanics. Even within mechanics things are not uniform: the characterization of stability used in statics is not the same as that used in dynamics.

First, let us discuss some statics problems. When we take a linear mechanical system and increase some load on the system in proportion to a parameter, we observe a proportional change in the solutions to the governing equations. Hence, for a linear system in such circumstances, we never see a sudden change in behavior. But everyday experience shows that many systems do not behave in this way. If we press down on an ordinary jar lid, it will yield slightly; with a bit more pressure, it will suddenly buckle inward. The resulting dimple often persists even after we remove our finger altogether.

Another interesting experiment can be done with a long, straight ruler. We compress it along its axis (assuming the ruler is straight, we should be able to apply a wide range of compression forces). In this case, there is a critical value of the compression force. Below this value, any small transverse deflection will disappear after removal of a transverse applied force. Above the critical value, however, the ruler will *buckle*; any transverse deflection force will lead to a transverse deformation that will persist until the axial compression is decreased below the critical value. So, at the critical value of the compression force, there is a loss of stability. We may speak similarly of a critical pressure value in relation to the jar lid.

In general, a system will change its behavior in some qualitative way when subjected to a critical load. This is what we called an "irregularity." As mentioned above, we must look to nonlinear systems for examples.

In dynamics, we have the problem of turbulent flow. Low-velocity pipe flow tends to be *laminar*: it is "quiet" and orderly, with fluid particles moving parallel to the pipe axis. At some higher velocity, the flow will become chaotic, with individual fluid particles moving unpredictably (even though the average motion will remain predictable). As a result, the flow resistance will increase and much more work will be necessary to transport the fluid. An understanding of *turbulent flow* is therefore of great practical importance. The transition between these two qualitatively distinct flow types is another example of a loss of stability.

Other examples can be found in ecology. In one such case, the dynamics of certain animal populations have been found to display a critical dependence upon pollutants in the environment.

We have associated the notion of instability with a qualitative change in system behavior. But the particular ways in which the subject of stability is treated varies markedly across disciplines. In statics, the subject is studied using, of course, the tools of statics. Here, a logical difficulty arises: a disturbance must be introduced in order to investigate stability, but any disturbance will result in at least some motion. So we are forced to speak in terms of motion, but restrict ourselves to static considerations (this approach is typical of thermodynamics as well). The tools for the investigation of time-varying problems are quite different and stand in better agreement with common sense. It is important that the static and dynamic methods of investigating stability often give rise to related results.

It is worth mentioning that although mechanical engineers are interested in stresses, strains, and the stability of structures, the bulk of modern mechanics is, as a science, centered more on singular values of parameters that describe processes. So stability is rather central to those working at levels more fundamental than that of common engineering design. The same thing can be said of physics as a whole: much of the modern parts of traditional physics serve to treat the irregular behaviors of objects.

Finally, we note that the crucial part of stability theory is *the particular definition of stability used* in the situation of interest. The results obtained can vary greatly when different definitions are employed. According to our fuzzy definition, loss of stability occurs at the moment when behavioral irregularities appear. For an ordinary object whose behavior is described by a usual function, points of irregularity could correspond to extrema of the function. We know that to locate extrema we must first find candidates for the extreme points and then test them. It would not make sense for us to expect more from the general problem of stability. So in order to define the terms "stability" and "loss of stability," we should

1. elaborate criteria for the loss of stability,
2. find out how to evaluate whether loss of stability actually occurs at a point that we have previously identified as a candidate for this, and
3. investigate the system behavior after loss of stability occurs.

These are the main aspects of stability theory.

3.16 Static Stability of a Straight Beam

A stability problem we can consider in some detail is one that was originally solved by Euler himself. This is a more serious version of the ruler problem posed in the previous section. The principles used in designing structural elements for bridges and buildings differ, in many cases, from those implemented by Nature. When a portion of a tree is damaged, we are likely to see a redistribution of strains in such a way that the organism can continue to function. Man-made structures, on the other hand, often operate at full capacity and with little safety margin. All too often a complete structural catastrophe ensues when buckling occurs somewhere in an important beam.

Let us consider a straight beam maintained under a longitudinal compression force P. The scheme for clamping and force application is shown in Figure 3.8. How large must we make P in order to produce the buckling phenomenon? Common sense tells us that at least one solution exists when the straight beam is compressed uniformly. We assume the existence of a critical value of P: when P just exceeds this value, the beam shows a small deflection that we recognize as the appearance of buckling. This solution

Figure 3.8 Geometry and mechanical configuration for beam stability problem.

is, however, close to the solution that gives a straight beam. From a mathematical viewpoint, then, the critical compression is a *point of branching* of the solution: before reaching the critical compression, we see only one solution, whereas after reaching it, we see at least two (both of which should depend continuously on P). This sort of branching behavior provides us with another way to identify points at which loss of stability might occur.

We have said that our investigations must be based on the nonlinear equations of bending. Because these are hard to solve, it is convenient to reduce the problem to a linear problem that retains the essential features. The idea of linearization of the initially nonlinear problem for this purpose was introduced by Euler. Above the critical compression we expect to have two solutions (i.e., deflection functions for the neutral axis of the beam): one, say, w_1, that is identically zero, and another, say, w_2, that differs slightly from w_1 (together with its derivatives). This prompts us to derive a linear equation for the difference $\Delta w = w_2 - w_1 = w_2$ (which can be arbitrarily small) from the nonlinear equation. We do this by linearizing, with respect to Δw, about the main solution w_1. Such a derivation, however, would be cumbersome. In the strength of materials, an equation for Δw can be derived directly through the introduction of some rather evident looking assumptions.[7]

Let us redenote the deflection Δw (which is small and occurring close to the critical point) by w. We assume that the resulting compression force at each cross section of the beam after deflection remains equal to P and parallel to the x-axis (the same assumptions we would make when linearizing the nonlinear equations with respect to Δw). We know that a couple M at a point implies a change in curvature of the neutral axis; for small deflections, this is

$$M = EI\frac{d^2w}{dx^2}.$$

We believe this formula applies to the compressed beam as well. By the above assumption, we see that the couple from the compression force (this is when we cut off one part of the beam and apply the force P instead at a new position of the neutral axis) is

$$M = -Pw.$$

---

[7]The wonderful thing about many such assumptions is that they can be rigorously justified. Indeed, the intuitive thinking of the early workers was often surprisingly accurate, and it played an important role in addressing complex questions.

Thus, the equation for w is

$$EI\frac{d^2w}{dx^2} = -Pw.$$

When EI is constant, we can take two derivatives[8] and get

$$EI\frac{d^4w}{dx^4} = -P\frac{d^2w}{dx^2}. \tag{3.26}$$

This is what we wished to obtain. The equation should be supplemented with boundary conditions, which for the pinned ends shown in Figure 3.8 are

$$w = 0, \qquad \frac{d^2w}{dx^2} = 0 \qquad \text{when } x = 0 \text{ and } x = L. \tag{3.27}$$

We obtained a problem that is linear in w. If we consider the solution w depending on the input P, however, then we have a nonlinear dependence $w = w(P)$.

The problem (3.26)–(3.27) with respect to w belongs to the class of *eigenvalue problems*. It has a trivial solution $w = 0$, but we wish to find nontrivial solutions $w \neq 0$ that correspond to some values of P. Let us find such a solution. Let $k^2 = P/EI$. Equation (3.26) has a general solution

$$w = c_1 \sin kx + c_2 \cos kx + c_3 x + c_4.$$

Conditions (3.27) imply that

$$c_2 = 0, \qquad c_3 = 0, \qquad c_4 = 0,$$

and

$$\sin kL = 0.$$

Thus

$$kL = n\pi,$$

where n is any integer, or

$$P = \frac{n^2\pi^2 EI}{L^2}. \tag{3.28}$$

For these particular values of P there exist nontrivial solutions $w = c_1 \sin kx$. The smallest value P_E occurs when $n = 1$, and it is called the *Euler critical*

---

[8]We take two additional derivatives in order to get the equation to agree with the general equation for the bending of beams, for which we can satisfy two boundary conditions at each end.

value; it was first found by Euler, who developed this method of linearization to find the critical points of branching of the dependence $w = w(P)$.

Thus, we have found the critical value of P. But c_1 can only be determined from a treatment of the full nonlinear equation (the derivation of which we avoided). Of course, we have obtained only a linear approximation to the dependence of w on P, so it holds only for small buckling deflections of the beam. If we wanted to find a higher approximation, we would have to apply the theory of the branching of solutions to a nonlinear boundary value problem.

There remains the question of whether buckling really happens as soon as the compression exceeds Euler's critical value. We know that the minimum value of system energy corresponds to a stable equilibrium point. It seems trustworthy that the lower the energy level of a configuration, the more probable it is for that state to be realized. Nonlinear analysis shows that the buckled state gives a lower level of energy in comparison with that for the straight compressed beam, so we conclude that this form is the one that occurs in reality for $P > P_E$. The problem solved by Euler is the basis for many other problems of stability. It turns out that exact experimental verification for the critical value P_E is hard to obtain because this value depends on (1) boundary conditions that are not so easy to implement exactly, and (2) the ideality of all the other conditions we introduced when we derived the problem.

We have obtained the Euler critical compression and said that it was what we needed. Such statements are common in cookbooks, but mathematics requires a proof. Let us therefore provide one. Of course, we shall not attempt to treat the full-blown nonlinear problem; however, in cases like this, a small step toward the complete proof can be valuable in its own right.

Let us show that for $P < P_E$ there is a unique solution $w = 0$ to the problem (3.26)–(3.27). Suppose that w satisfies (3.26)–(3.27). Multiplying (3.26) by w termwise and integrating over the beam, we get

$$\int_0^L \left(EI\frac{d^4w}{dx^4} + P\frac{d^2w}{dx^2} \right) w\,dx = 0.$$

Integrating by parts in both terms (twice in the first one), we obtain

$$\int_0^L \left(EI\left(\frac{d^2w}{dx^2}\right)^2 - P\left(\frac{dw}{dx}\right)^2 \right) dx = 0. \tag{3.29}$$

Here, the boundary evaluation terms vanish by (3.27). From this equality, for $P < P_E$ it follows that $w = 0$.

The proof of this last statement will be sketchy, and a reader inexperienced with the calculus of variations can simply omit it without loss of continuity. Let us consider the following isoperimetric problem of the calculus of variations. On the set of all functions from $C^{(4)}(0, L)$ satisfying the conditions (3.27) and the constraint

$$\int_0^L \left(\frac{d^2 w}{dx^2} \right)^2 dx = 1,$$

we seek the one for which the integral (functional)

$$\int_0^L \left(\frac{dw}{dx} \right)^2 dx \tag{3.30}$$

attains its maximum. The solution is found by the method of Lagrange multipliers: the required function is an extremal of the functional

$$\int_0^L \left(\frac{dw}{dx} \right)^2 dx - \lambda \int_0^L \left(\frac{d^2 w}{dx^2} \right)^2 dx.$$

An extremal of this functional satisfies the so-called Euler equation, which is

$$\lambda \frac{d^4 w}{dx^4} + \frac{d^2 w}{dx^2} = 0.$$

This coincides with (3.26) up to the notation $\lambda = EI/P$. Here we need nontrivial solutions, so everything we said above about the eigenvalue problem (3.26)–(3.27) can be reformulated for this problem, and the greatest value for (3.30) occurs for the greatest value of λ because a solution of the problem corresponding to λ will satisfy, similar to the above, the equation

$$\int_0^L \left(\lambda \left(\frac{d^2 w}{dx^2} \right)^2 - \left(\frac{dw}{dx} \right)^2 \right) dx = 0.$$

This greatest λ, for which there is a nontrivial solution to the Euler equation with boundary conditions (3.27), corresponds to the least P for which (3.26)–(3.27) has a nontrivial solution. So, for $P = P_E$, we have

$$\int_0^L EI \left(\frac{d^2 w}{dx^2} \right)^2 dx \geq P_E \int_0^L \left(\frac{dw}{dx} \right)^2 dx,$$

which holds for any function from $C^{(4)}(0, L)$ satisfying (3.27). Thus, for $P < P_E$, for a solution satisfying (3.29), we find that

$$\left(\frac{P_E}{P} - 1 \right) \int_0^L EI \left(\frac{d^2 w}{dx^2} \right)^2 \leq 0.$$

Therefore, $d^2 w/dx^2 = 0$, and because of the boundary conditions (3.27), we have $w = 0$ as desired.

When $P > P_E$, we cannot show nonuniqueness of solution to the problem (3.26)–(3.27) if P is not an eigenvalue. To demonstrate that buckling becomes possible for any $P > P_E$, we need access to other tools. These will be given in the next section.

3.17 Dynamical Tools for Studying Stability

Our first inclination is to associate the stability of an object with the object's temporal behavior, so it makes sense to take the dynamical equations and examine how deflections develop over time. For the additional deflection w of a compressed beam, we have

$$\rho \frac{\partial^2 w}{\partial t^2} + EI \frac{\partial^4 w}{\partial x^4} + P \frac{\partial^2 w}{\partial x^2} = 0. \tag{3.31}$$

In the first inertia term, ρ is the linear density of the beam. This equation should be supplemented by boundary conditions (we shall assume pinned ends, as in the previous section) and initial conditions

$$w\big|_{t=0} = w_0(x), \qquad \frac{\partial w}{\partial t}\bigg|_{t=0} = w_1(x). \tag{3.32}$$

Common sense tells us that the notion of stability relates to the time behavior of perturbations: if any small perturbation yields only small solutions to the above problem, then the main solution is stable. If for some small perturbation the solution is not small, then the main state is unstable. A rigorous treatment would require us to define the term "small," of course, but we will leave this to more advanced books.

Later, we will show that for $P > P_E$, there is a small initial perturbation (w_0, w_1) such that w grows in an unbounded fashion over time. This means that the straight compressed state of the beam is unstable in the above sense. But what happens in dynamics to the stability of the straight state of the beam for $P < P_E$? Is it possible that the static stability of this state can change to instability in a dynamical sense? We need to investigate this, because the inclusion of inertia terms in the equations can change the solution properties in a crucial manner. We should not identify the properties of a real object with those of its mathematical model, so any property that seems evident from our everyday experience should be verified by mathematical tools!

We shall demonstrate that for $P < P_E$, the straight compressed state of the beam remains stable in the dynamical sense as well. In order to do this, we shall obtain an estimate of the solutions to (3.32) for the beam having pinned ends. This estimate follows from energy conservation for the beam. We shall derive the particular form of this that we need, and we will do so rather quickly — without showing too much detail.

We multiply the terms of (3.31) by $\partial w / \partial t$ and integrate with respect to x over $[0, L]$:

$$\int_0^L \rho \frac{\partial^2 w}{\partial t^2} \frac{\partial w}{\partial t} \, dx + EI \int_0^L \frac{\partial^4 w}{\partial x^4} \frac{\partial w}{\partial t} \, dx + P \int_0^L \frac{\partial^2 w}{\partial x^2} \frac{\partial w}{\partial t} \, dx = 0.$$

Integrating by parts in the second and third integrals, we get

$$\int_0^L \rho \frac{\partial^2 w}{\partial t^2} \frac{\partial w}{\partial t} \, dx + EI \int_0^L \frac{\partial^2 w}{\partial x^2} \frac{\partial^3 w}{\partial t \partial x^2} \, dx - P \int_0^L \frac{\partial w}{\partial x} \frac{\partial^2 w}{\partial t \partial x} \, dx = 0.$$

It is seen that this can be represented as

$$\frac{1}{2} \frac{\partial}{\partial t} \left(\rho \int_0^L \left(\frac{\partial w}{\partial t} \right)^2 \, dx + EI \int_0^L \left(\frac{\partial^2 w}{\partial x^2} \right)^2 \, dx - P \int_0^L \left(\frac{\partial w}{\partial x} \right)^2 \, dx \right) = 0.$$

Integrating both sides of this in t over $[0, \tau]$, we obtain

$$\left(\rho \int_0^L \left(\frac{\partial w}{\partial t} \right)^2 \, dx + EI \int_0^L \left(\frac{\partial^2 w}{\partial x^2} \right)^2 \, dx - P \int_0^L \left(\frac{\partial w}{\partial x} \right)^2 \, dx \right) \Bigg|_{t=\tau}$$

$$= \left(\rho \int_0^L \left(\frac{\partial w}{\partial t} \right)^2 \, dx + EI \int_0^L \left(\frac{\partial^2 w}{\partial x^2} \right)^2 \, dx - P \int_0^L \left(\frac{\partial w}{\partial x} \right)^2 \, dx \right) \Bigg|_{t=0}.$$

This is the equation we need.

Taking w_0, w_1 small enough, we get a small value ε for the right-hand side. Thus, the same value is attained by the left-hand side at any instant $t = \tau$:

$$\left(\rho \int_0^L \left(\frac{\partial w}{\partial t} \right)^2 \, dx + EI \int_0^L \left(\frac{\partial^2 w}{\partial x^2} \right)^2 \, dx - P \int_0^L \left(\frac{\partial w}{\partial x} \right)^2 \, dx \right) \Bigg|_{t=\tau} = \varepsilon.$$

It follows that w is small for $P < P_E$; let us demonstrate this next. We need the inequality

$$\int_0^L EI \left(\frac{d^2 w}{dx^2} \right)^2 \, dx \geq P_E \int_0^L \left(\frac{dw}{dx} \right)^2 \, dx$$

of the previous section. We have

$$\left(\rho \int_0^L \left(\frac{\partial w}{\partial t} \right)^2 \, dx + \left(1 - \frac{P}{P_E} \right) EI \int_0^L \left(\frac{\partial^2 w}{\partial x^2} \right)^2 \, dx \right) \Bigg|_{t=\tau} \leq \varepsilon.$$

We therefore see that the integrals

$$\int_0^L \left(\frac{\partial w}{\partial t} \right)^2 \, dx, \qquad \int_0^L \left(\frac{\partial^2 w}{\partial x^2} \right)^2 \, dx,$$

remain small enough at any instant. Of course, this is smallness in an average sense only, but for pinned boundary conditions, $w(x,t)$ and $\partial w(x,t)/\partial x$ can be estimated using these integrals.[9] So we can state that the compressed straight state of the beam is stable for $P < P_E$, as desired.

Whereas stability must be established with respect to all small perturbations, instability can be established for just one. Let us show that for $P > P_E$, the straight state is unstable. For this, we obtain a special solution to (3.31), related to the Euler buckling we found in our statical analysis. Namely, let us seek a solution of the form

$$A(t) \sin \frac{\pi x}{L},$$

with unknown magnitude $A(t)$. It is clear that this satisfies the pinned end conditions. Substituting this into (3.31), we get

$$\rho A''(t) \sin \frac{\pi x}{L} + EIA(t)\frac{\pi^4}{L^4} \sin \frac{\pi x}{L} - PA(t)\frac{\pi^2}{L^2} \sin \frac{\pi x}{L} = 0.$$

Canceling the factor $\sin(\pi x)/L$, we find that A must satisfy

$$\rho A''(t) + \left(EI\frac{\pi^4}{L^4} - P\frac{\pi^2}{L^2} \right) A(t) = 0.$$

This is of the form

$$A''(t) - k^2 A(t) = 0,$$

where the parameter k is positive when $P > P_E$. This equation has the general solution

$$A = c_1 e^{kt} + c_2 e^{-kt};$$

hence the solution to (3.31),

$$\eta e^{kt} \sin \frac{\pi x}{L},$$

grows without bound for any initial value η (no matter how small). So, we have completed the proof, and have thereby established the desired stability properties of the compressed straight state of the beam.

3.18 Additional Remarks on Stability

In the previous sections, we considered only one problem of stability for a mechanical system, and found it to be fairly challenging. To make matters

---

[9]This is done using the modern theory of Sobolev imbedding. However, in this case, the corresponding inequalities could be demonstrated using elementary tools.

worse, stability analysis plays an integral role in an extremely wide variety of mechanics problems. The single technique we considered certainly cannot cover all cases. The reader should be aware that nowadays the dynamics-based approach is thought to be the most reliable; in fact, it is the only possible approach for many problems. Our statical approach was based on the energy criterion for stability. When nonpotential forces act on a system, this approach is not applicable: this happens, for example, with a traction force whose line of direction depends on the change in shape of the object. In such cases, only dynamical considerations can answer questions about stability. However, the static tools for investigating stability were extended to problems involving not only changes in mechanical characteristics but also changes in temperature. Gibbs found four important thermodynamical processes for which stability can be investigated by tools similar to the energy approach in statics. For this, it was necessary to compose the potentials for a system. Gibbs found four types of potentials, each corresponding to a process amenable to the static approach. One of these processes is adiabatic. We will not pursue this further, but the reader can consult a textbook on thermodynamics for more details.

We would like to touch on the question of how loss of stability occurs. Some mechanical systems undergo this in a more or less subtle manner: a small increase in load gives a small deviation from the main unstable state. But we cited the example of a jar lid that can abruptly snap over to a new state of curvature when we press on it; furthermore, the new state may persist after removal of the applied force. In this case, it is possible to speak of a *catastrophe*. While this may seem to be an odd term for a jar lid's state, we may see essentially the same effect on the roof of a building, or on the metallic surface of an airplane or submarine. Then disaster is a real possibility. Loss of stability can bring on a crucial change in system behavior in many different systems, both natural and man-made (because many systems are described by rather similar mathematical models). The relevant problems from various areas of study have been collected under the name *catastrophe theory*. This science has spawned a terminology and has taken on a life of its own. Although it has not been able to enhance our understanding of many of the older problems of mechanics, it is useful in the way it serves to unify problems from many different areas.

We can extend the dynamical approach to the study of stability, and thereby treat other problems. An old stability problem is concerned with the motion of the Earth. The Earth revolves around the Sun, and the Moon revolves around the Earth. Other large bodies (planets, asteroids, etc.) are also present. We know this system has existed for a long time. Although things may not change much in our lifetimes, what will happen to the motion of the Earth in the distant future? The simplest problem to consider is the nature of this motion as time $t \to \infty$.[10] People have always

---

[10]Of course, we have thereby left the realm of practicality: the Sun, although it will outlive us, will certainly not exist forever. But such abstract problems are often easier

wondered whether the orbital motion of the Earth is stable or whether it would someday change in such a way as to make life impossible.

Such motions are described by ordinary differential equations. We normally wish to learn how a solution to a corresponding Cauchy problem depends on slight changes in external (including initial) conditions over the time period $[0, \infty)$. We would call the solution stable if a small change in the external conditions implies a small change in the solution. The data considered as external for classical mechanics problems are the forces, masses, and initial conditions; however, in other branches of science, the external data could be of a different nature. The notion of stability normally employed is called *Lyapunov stability*.[11] This characterizes the condition of stability with respect to disturbances in the initial data. To understand why the formulation is so restricted, let us consider an elementary problem. A simple oscillation system consists of a mass m attached to a spring of rigidity k and subjected to a force $F(t)$:

$$mx''(t) + kx(t) = F(t).$$

When $F(t) = 0$, the general solution is

$$x = A\sin(pt + \phi), \qquad p = \sqrt{\frac{k}{m}},$$

with an arbitrary amplitude A and phase ϕ for the oscillations. Any solution is evidently stable with respect to disturbances in the initial data for x, but if we add a small force $F(t) = \varepsilon \sin pt$, we encounter oscillations that increase in amplitude without bound. This is the familiar effect known as *resonance*. From this viewpoint, then, the free oscillations of the system are unstable. A small periodic force can have such a large cumulative effect because during each period it increases the energy of the system.

However, in many circumstances we can neglect time-changing forces and consider only undriven oscillations. The Sun-Moon-Earth system is an example of this; the net influence of other planets and bodies is negligible. In such cases, we can limit ourselves to stability considerations predicated on changes in the initial data only. The definition of Lyapunov stability is somewhat similar to that of continuity.

Definition 3.1 Consider a system of ordinary differential equations

$$\mathbf{x}'(t) = \mathbf{f}(\mathbf{x}(t), t)$$

with initial condition $\mathbf{x}(t_0) = \mathbf{x}_0$. A solution $\mathbf{x}(t)$ on the interval $[0, \infty)$ is *stable in the sense of Lyapunov* if for any $\varepsilon > 0$, there is a $\delta > 0$, such that

---

to solve, and this explains why many problems that could be formulated for a finite time period are formulated for an infinite time period instead.

[11] Aleksandr Mikhailovich Lyapunov (1857–1918).

for any $\mathbf{x}_1$ from the δ-neighborhood of $\mathbf{x}_0$ (i.e., such that $\|\mathbf{x}_1 - \mathbf{x}_0\| < \delta$), the solution $\mathbf{y}(t)$ to the same equation but with the initial data $\mathbf{x}_1$ satisfies $\|\mathbf{y}(t) - \mathbf{x}(t)\| < \varepsilon$ for all $t \in [t_0, \infty)$. Roughly speaking, this means that any trajectory that starts close to a stable trajectory will stay close to it for all time. If in addition we have $\|\mathbf{y}(t) - \mathbf{x}(t)\| \to 0$ as $t \to \infty$ for any such $\mathbf{y}(t)$, the solution $\mathbf{x}(t)$ is said to be *asymptotically stable in the sense of Lyapunov*. So, an asymptotically stable trajectory is stable, and any trajectory that starts close to it will approach it more and more closely as time goes on.

This is the main definition used to study the stability of motion, and has led to success with a great many problems (not including, however, the problem of the Earth mentioned above). In particular, the notion of asymptotic stability is useful in control theory: a typical control system has to actually restore the original system state after a perturbation has occurred.

Because Lyapunov's notion of stability cannot cover all instances, other definitions have arisen. We could consider a mass revolving around another, much larger mass. A slight change in the smaller mass will change its period of revolution. This means that if the initial and perturbed systems are compared, then the distance between the initial and perturbed masses can become significant over time, although the orbital shapes will remain close. (Similar things can happen with nonlinear oscillations.) So it makes sense to introduce a notion called "orbital stability of motion." We could give other examples, but will leave the interested reader to consult more specialized books. There, one can also learn about Lyapunov's two methods for studying the stability of various motions in the Lyapunov sense.

Many seemingly simple problems of real life are complex when considered mathematically. As an example of how ordinary circumstances can lead us into difficult analysis (and high technology at the same time), we consider the problem of leaks.

3.19 Leak Prevention

Many people are unaware of the tremendous effort that goes into making everyday items leakproof. A leaky roof can, of course, be a total disaster. This holds true even more so for a ship's hull. Any perforation in an airplane, rocket, or submarine must be closed carefully.

The problem of leakproofing various connections and couplings is especially important. A generous water leak on the ninth floor of a building can cause problems all the way to the basement. Military jets are often refueled in midair; even a minor leak can result in the demise of both aircraft involved. Flange-type couplings are ubiquitous in systems designed to transport liquids: water, oil, gas, and so on. We find these in car engines, hot water heaters, oil tankers, and numerous other places. The average person will simply not appreciate the time, effort, and money that must be put into the goal of making things leakproof: the total expenditure has far exceeded that required to develop even space-age items such as nuclear weapons. The amount of research is not readily apparent because each large company has its own research department performing tests on pipes, hermetic seals, and so forth. Carefully guarded as trade secrets, few of the results are ever published in journals. Sometimes the success or failure of an entire project will hinge on an ability to produce a reliable hermetic joint for use at high temperatures or pressures. (The types of joints used in an ordinary house, though obviously subject to much less severe operating conditions, took generations of engineers to develop. The situation was similar with many things we now take for granted.) The problem is crucial when volatile chemicals are involved: thousands of square kilometers of land and water have been polluted by oil spills.

A flange joint is a simple structure. We have all seen these on large water pipes: the two pipes to be joined each have a hard ring attached at the end, and bolts serve to tighten these rings down on a gasket seal sandwiched in between. The gasket is typically made of resin or a synthetic material, but sometimes of metal. A surprising number of engineering mistakes were committed in the development of this simple thing, however. It turns out that the gasket behaves in a highly nonlinear fashion when compressed. In addition, friction between the gasket and solid portions of the flange becomes an important consideration.[12] Indeed, a thorough analysis of a simple flange joint would pose a challenge for the most well equipped mechanicist. The pipes, rings, and bolts themselves behave more or less linearly and are subject to rigorous design methods. However, nonuniform temperature distributions remain a source of difficulty even for flanges used in ordinary civil engineering structures. For the gasket itself, engineering textbooks provide the simplest guidelines for sizing to preclude leaks under certain pipe pressures. These guidelines are purely empirical in nature, derived from trial and error over many years of practice. However, with resin-type materials, some tricky things can happen. Although we tend to think of resins as being somewhat "soft" (because of the way they behave under tension), they can provide steel-like resistance under forces of high volume-compression — their range of thickness accommodation is therefore tiny. It turns out that the flange rings in contact with the gasket can undergo significant twisting motions when subjected to normal temperature gradients (say, between the liquid under transport and the external environment). Therefore it is hard to keep a flange joint tight throughout all seasons of the year. Things can get especially bad if the designer fails to take into account the difference between

---

[12]We have not discussed the problem of friction in this book, but it gives rise to a complex theory known as *tribology*. The presence of friction is a source of great complication for problems in elasticity. Engineers can introduce various "coefficients of friction," but these merely provide bounds on the friction forces. Sometimes simply touching a surface with one's fingers can significantly alter the amount of friction present because of lubrication. The theory of lubrication is still somewhat of an art form, and practical work is often accomplished on the basis of intuition.

the environmental conditions under which the joint is to be assembled and those under which it is to be deployed.

At first glance, such problems seem to belong purely to engineering. However, they provide a real challenge for theoreticians as well: they encourage the development of new models in continuum mechanics. Every science goes through a stage in which it seems that only empirical approaches are possible. Persons eventually come along who are able to create real science on the basis of the empirical information that has been amassed.

Chapter Four

Some Questions of Modeling in the Natural Sciences

4.1 Modeling and Simulation

The term "modeling" is very broad. It sometimes suggests the production of a scale model of a real object using inexpensive materials. In mathematical modeling, on the other hand, we compose a mathematical description of an object and then try to learn how the real object behaves in various circumstances. However, even modeling with the use of materials — like the use of a small model of an airplane in an airstream — requires mathematical modeling, because the interpretation of the results of direct physical modeling is not straightforward: when one changes the dimensions or materials of the object, it is impossible to provide full similarity in behavior between the real object and its model (the changes in linear dimensions, areas, and volumes, as well as forces of interaction, are not proportional). So the use of physical models requires a mathematical accompaniment in any case. Mathematical simulation has become the main tool of the physical sciences, but things are moving in the same direction in many other areas, for example, biology, linguistics, and economics.

If we wish to apply mathematics in any science, we must find a way to characterize the properties of the object under consideration in numerical form. For some subject areas, this may seem impossible at first glance, but the analysis of a given object will often serve to identify a property that can be characterized by its intensity, or something similar. A practitioner is then on the way toward the use of mathematics to represent, analyze, and combine complex objects. One such unexpected application of mathematics to a nonmathematical area is seen in a humorous theorem attributed to Lev Landau (1908–1968), a Nobel Prize winner in physics:

> For any woman there is a distance at which the woman looks most attractive and sympathetic.

The "proof" resembles that of a calculus theorem. We cannot see an object if its distance is zero or infinite, so its attractiveness at these distances is zero. Because the attractiveness is a continuous function of the distance, there must be an intermediate point at which it takes on its maximum value (and, unfortunately, another point at which it takes on its minimum).[1] Landau's

---

[1] Here, we have introduced attractiveness as a continuous function of the distance. It seems that we cannot apply the theorem on the maximum of a continuous function on a

little theorem suggests that it is possible to "measure" many diverse phenomena, even those (like feelings) for which the application of mathematical tools may seem inappropriate at first.

The prevalence of computers will introduce mathematics into all areas of human activity. In support of this is the fact that our world is composed of atoms or molecules whose motions can be described (at least to some degree of approximation) mathematically.

At one time, mathematical models had to be kept simple because people could solve only algebraic or differential equations. People working in physics, in particular, were hindered by having to work with short, simple equations. As soon as these limitations lessened, it became possible to employ huge systems of simultaneous equations, inequalities, and other mathematical objects to describe real phenomena. At present, discrete approximations of models can contain tens of thousands of algebraic equations, and nobody cares how nice they look. (However, when they do look nice it gives great satisfaction to the researcher, who can then understand something about what the computer is doing.) The mathematical modeling of real-life objects requires the development of various mathematical tools in addition to those of a strictly computational nature.

The closest connection that mathematics has with a natural science is the connection it has with physics and engineering. So modelers working in other sciences should attempt to learn something from the interaction between these sciences and mathematics.

In research there are various approaches for the formulation and use of models. These approaches could be characterized, respectively, as those of a mathematician, an engineer, and a physicist. The first approach is that of a user who believes his model is ideal and should not (or cannot) be changed. This is the viewpoint of the mathematician who has obtained a ready-made mathematical model of something. He never tries to change anything, but seeks only the properties of the equation he has received from an engineer or physicist. He solves equations, proves theorems, and so on. In other words, a mathematician uses the axiomatic approach: for him, the model is something like an axiom. It is unfortunate that when he runs into something he cannot prove or solve, he merely continues his attempts to the point of frustration.

The two other approaches concern how to formulate models and use them. An engineer, when she finds she cannot solve something or that a model does not fit the expected properties of an object, says, in effect, "I need to change the main assumptions I introduced in deriving these equations." In this way, she obtains equations that can be solved or that possess needed properties.

The approach of a typical physicist is slightly different. He may say to himself, "We know little about the object under consideration. I cannot solve these equations directly, but if I add a certain small term to the equations

---

compact interval, because $[0, \infty)$ is not compact; however, it is possible to "correct" the proof through the introduction of a metric under which this semi-infinite interval becomes compact.

or change that term a bit, then I can solve the modified equations nicely. At this point, I do not know what this would mean physically, but the slightly disturbed equations should yield a good approximation. After I solve the problem, I will consider ways in which I could interpret the perturbation I have introduced." Sometimes such a perturbation is senseless and cannot be interpreted physically. However, it often turns out that such changes really have a physical meaning, because the physicist has experience and some fuzzy ideas regarding how to disturb the equations to get a result. His ideas are based on some understanding of the problem, and thus really do reflect the properties of the object under consideration.

Engineers and physicists normally believe that the work of mathematicians on a model is purely technical and serves only to muddy the waters around an otherwise clear question. Mathematicians are regarded as useful only when something goes wrong with the details of calculations. The truth is, however, that all these players are studying the same problem using different tools. Their viewpoints differ, as do their understandings of the goal. But the problem is the same and they attack it from various sides, continuing to take bites out of it until it begins to yield as a whole; so their collaboration in the process is extremely important.

4.2 Computerization and Modeling

We attempt to simulate real-world processes through the use of mathematical models. Here, it is important to understand that the results we obtain will reflect not only the processes themselves, but our own limitations as well. Problems that challenged mechanicists two centuries ago are now considered easy, and in another century our present research problems will be viewed similarly. The tools available to any given generation of investigators will determine both the accuracy they can obtain and the complexity of the models they can employ.

Early models in continuum mechanics had to be kept comparatively simple. The equations used were mostly linear. Civil engineering problems were solved using graphical methods; these were essentially boundary value problems for ordinary differential equations, and yielded to fairly simple calculations involving given forces and moments. More complex problems had to be described using partial differential equations. Those within reach involved comparatively simple domains such as rectangles, circles, and balls. To attack these problems, various analytical approaches were developed.

The use of computers has revolutionized mechanics. Problems that challenged experts fifty years ago are now assigned to students as simple exercises, while design engineers rely on standard computer software to accomplish their tasks (often without a full understanding of how the programs actually work). Software viability can be tested on some of the more difficult elasticity problems; otherwise, interest in these seems to have become largely historical.

Modern computers have given us the ability to solve nonlinear problems of ever increasing complexity. This, of course, means that the complexity of the models used to describe practical objects will continue to increase. For the most part, the old approximate linear theories now find their uses in engineering design at elementary levels.

The assertion is commonly heard that computers can solve all the problems of mechanics, and that traditional mechanics is dying. This is roughly equivalent to the opinion of a child who has learned the alphabet and concludes that he or she knows the entire literature of mankind. It is true that powerful computers have radically increased our abilities to attack certain types of problems; correspondingly, as we stated above, interest in some classes of problems has been transferred to the realm of low-level engineering design. But we are encountering new problems and regimes in mechanical research where the attack must proceed through the simultaneous use of computers and analytical approaches. At this point, it is still true that the main tool of research is the human brain. One can only hope that this situation will persist for at least as long as there are persons motivated by curiosity as opposed to base utilitarian interests. Mechanics provides us not only with a simple record of the facts and methods used to calculate forces, stresses, and displacements; it offers us new ideas about relations in nature and about the objects produced by human civilization. We have the chance for a much deeper appreciation of the world around us. The results of work on new mathematical models can serve to heighten what we often call our "intuition" about how things work. This is the prized possession of any real researcher, and gives him or her the chance for true aesthetic expression. As time goes on, we should expect to see more and more routine tasks assigned to computers. But researchers are still charged with understanding the results, and with developing new ideas and theoretical statements without which the work of computers will be senseless. Computers cannot originate ideas. Being finite automata, they can outperform the human brain in the arena of simple operations. They cannot stray outside the framework of the code under which they operate, however, and that code is still written by human beings.

Science fiction movies have introduced society to the notion that computers will fulfill all the duties of people, the only danger being that computers might become independent and try to dominate their creators. But computers suffer from at least two shortcomings when compared to the human brain. First, they lack real understanding. Computer codes for doing calculations, for example, are so general that they suggest a large set of potential models for use — intelligent selection must therefore be done by a human being. The human touch is also needed for the development of new ideas lying at the base of particular mathematical models, and for the proposal of the best algorithms for calculation. Second, finite automata can only deal with finite number representations and with finitely many operations; this means that some inaccuracy will always be inherent in the work of computers. A researcher must become proficient in the art of controlling the effects of these numerical errors; often he or she will have to rewrite a portion of the computer code. This is where one can get into trouble with "black box" codes. It is dangerous to think that one can supply a routine with input data and then rely on the output data without knowing exactly how it was obtained. This is why most proficient researchers will choose to write their own computer codes. A code tailored to a certain application will lack generality, but may vastly outperform an "all-purpose" code. This situation is likely to persist for a long time. And again, the composition of such codes requires a real (i.e., human) understanding of the subject.

Finally, we might say that the impressive-looking output we see coming from the use of computers represents no more than a kind of still photograph of reality. For millenia, mankind has watched many natural processes, but their explanation and use has come only after the development of the special tools that have accompanied the rise of science. An understanding of the nature of an object or process comes after a theoretical analysis based on knowledge of the main principles of a science, taking into account both the boundaries of applicability of the models and the limitations inherent in the numerical methods employed. It is almost an axiom that an ignorant person cannot use complex mathematical software. Even those programs that offer us the chance to "push a button and see" require an observer who can properly see. Besides, there will always be problems that do not fall under a common scheme, and thus a need for persons who can prepare the way for others to simply push a button.

We should also stress that new capabilities in calculation will necessarily bring new viewpoints on what constitutes a good model in mechanics and the other sciences. Perhaps continuum mechanics will be revised in order to more closely embrace the ideas of atomic physics. Without doubt the priorities of many applied sciences like mechanics will change as the solution of many problems becomes routine work. The appearance of new materials and nonclassical problems always poses challenges — first centered around modeling, then around constructing reliable solution methods. The adequacy of a new model must always be assessed on both theoretical and experimental grounds; its properties have to be investigated on both quantitative and qualitative bases. The human brain will play an essential role in these activities for a long time to come.[2]

---

[2]This is not to say that the present high demand for computer programmers will persist. Indeed, we can expect to see more *routine* programming tasks relegated to computers themselves. We may ultimately reach a situation in which the only human computer programmers required will be specialized experts in certain fields of knowledge or application. These persons will probably serve to formulate computer tasks in the terms to which people working in the area are accustomed. A certain number of people will be required to work in the theory of computation, but probably not many.

4.3 Numerical Methods and Modeling in Mechanics

When deriving a partial differential equation (PDE) to describe an object of continuum mechanics, we completely ignore structure at the atomic level. During the derivation, we undertake limit passages requiring the material to be more or less homogeneous at the point of interest. We cannot perform a true limit passage while considering atomic structure; we find we must stop reducing the relevant distances as soon as they become smaller than several atomic diameters. So the models of continuum mechanics really serve to describe a sort of ideal continuous material with no atomic structure. Hence, they are imprecise descriptions, even if we make no simplifying assumptions regarding linearity, and so forth.

The boundary value problems of mechanics are difficult. Most cannot be solved analytically, but modern computers allow us to employ numerical tools. By their very nature, these tools are always approximate: regardless of the particular method of discretization we employ, we are invariably led to a finite (although sometimes large) number of simultaneous algebraic equations. Such a system can be regarded as a finite-dimensional model of the object described by the original PDE. It may be better considered as a model of the PDE itself because, as we noted above, the latter may not precisely represent the mechanical object in the first place.

The fact that some equations in the complete setup of a boundary value problem may be "precise" (because they follow from the laws of continuum mechanics) does not imply that the continuum model is "better" than the same model after discretization. The "goodness" criterion for many models is not the accuracy with which the boundary value problem is approximated, but rather the accuracy with which real-world quantities are approximated. This is the viewpoint of the design engineer who deals with actual experimental data. The opinion of a pure mathematician, of course, may be just the opposite.

In support of the engineering side we offer the following example. Suppose the solution of an elasticity problem shows that at some point the stresses become infinitely large (this is commonly seen with solutions of contact problems near the boundary of contact). We know that in a real body this cannot occur: in steel, for example, the stresses are automatically redistributed via plasticity before they can become too large. A designer will rightly conclude that the model of linear elasticity has failed at such a point. Why then should he or she require a finite numerical model to exhibit precisely the same behavior? A knowledge that the finite model provides a good approximation away from a few singular points may be enough. There is no practical need to prove that an increase in calculation accuracy will yield results closer and closer to the solution of some partial differential equation.

Nine out of ten engineers consider the rigorous proofs offered by mathematicians as unnecessary, as long as satisfactory numerical results can be obtained without rigor. They do, however, accept it as useful when mathematicians can point out cases in which solutions do not exist or cannot be accurately approximated in principle.

So, we might come to believe that the majority of practitioners have the right viewpoint; however, this is true only to some extent. First, we can never obtain sufficient data about an object to assert that a given model is adequate under all circumstances. A model will be able to generate predictions about what is happening at points that fall outside the reach of measuring instruments. Next, the process of discretization can strip away important physical properties and may even lead to contradictions with established physical principles. This explains why mathematicians take the first but necessary step of studying whether numerical solutions really represent the solutions to initial and boundary value problems. It is also important to demonstrate the "well-posedness" of a numerical method, despite the fact that computer solutions cannot converge to the precise ones: because of truncation, only those numerical solutions done to a certain level of approximation can come close to the actual solutions of boundary value problems. It is also true that a continued increase in the number of approximating equations will eventually cause the approximate solutions to deviate from the "precise" one. Nonetheless, a theoretical check on a method through the use of a limit passage can serve as an important indicator of the method's potential for accuracy.

We should add, however, that some discrete models (such as those obtained through various versions of the finite element method) can be of real interest even when we cannot justify convergence of their approximations to the exact solution (as is the case for many nonlinear problems). In such instances the investigation of a finite-dimensional model, without reference to a boundary value problem for a PDE, can have an important meaning.

Let us also note that some models of continuum mechanics — such as models of beams, plates, and shells — can be regarded as versions of the finite element method applied to multidimensional problems: we prescribe a certain behavior (strain and stress) for elementary portions of these structures in exactly the same way as we do for the classical finite element method (where we specify how a finite element can behave). So, when someone announces that it is time to discard all simple approximate theories and rely on three-dimensional computer solutions instead, let us remind them that the assumptions behind many approximate theories are so good that these theories can easily outperform any robotlike application of the finite element method.

In the opinion of a mathematician, what are the necessary steps for investigating a typical boundary value problem along with the possibilities for its numerical solution? The first thing to note is that once a problem was formulated, it acquired a life of its own (independent of its industrial or natural origin). It is illogical to use the properties of some natural object to argue that a model for that object behaves in a certain way. Such properties can only serve as a guide — a source of suggestions for investigation. When we find that a model does display a desired property, we can become no more than encouraged. *The object and its model are independent*; this is the key statement of mathematical investigation. So a mathematician will first ask about well-posedness, and here there will be three main issues.

1. *Existence of solution.* An attempt will be made to establish a theorem giving conditions either necessary or sufficient to guarantee that the problem is solvable in principle.
2. *Uniqueness of solution.* Is there only one solution, or are multiple solutions possible (depending upon the values chosen for some parameter)?
3. *Qualitative behavior of solutions.* Here, we are interested in the dependence of solutions on external parameters.

He or she may then proceed to justify some numerical approach to the problem. Will the discretized equations be solvable (and, if so, will the solution be unique)? Will successive approximations converge to the exact solution of the boundary value problem? How fast will this convergence occur, and will it be possible to accelerate it? Often only some of these questions can be answered with the tools available at a given time. Nonetheless, an increased understanding is better than none at all. Sometimes many of the steps we have mentioned can be taken in parallel; this is the case when problems are approached using variational numerical methods.

4.4 Complexity in the Real World

Once we come to understand something, we tend to think of it as simple. Other things appear quite complex. Before Hooke's time, for example, elastic behavior could not be regarded as simple. The same holds for many other effects and scientific notions. In science, we have become accustomed to taking an object or process and selecting only certain of its main features for study; we effectively pretend it has no other features or properties. This is how science originated, and it remains the way in which complex mathematical models are elaborated. But our choice of "main" features can be misguided if based on everyday experience. Indeed, the complexity of many events turns out to far exceed our initial expectations. We would like to offer a few examples.

Every car has a gear box that serves to transfer the engine power to the wheels. The gears stand in an appropriate ratio, with smaller ones driving larger ones to yield a certain mechanical advantage. A naval destroyer will also have a gear box. In this case, however, the gears can exceed 10 meters in diameter and their teeth can reach a linear speed of 100 meters per second. The box must be efficient and able to function for long periods without maintenance, all the while transferring an enormous amount of energy to the ship's propeller via a relatively tiny area of mechanical contact (between the gear teeth). A statical analysis would yield a contact pressure so large that the gears should fail immediately. But upon opening a real box ten years down the road, one can discover gears in a "like new" condition. This results from good design: metal-on-metal contact is avoided through the inclusion of an oil layer between the gear surfaces. At high velocities, this oil behaves in ways other than those we might anticipate based on our everyday experience. An appropriate model must therefore incorporate the dynamics (and possibly chemistry) of a viscous liquid into a dynamical problem of elasticity. Here, everything will be important: the gear shape, type of oil, and so on. The difficulty of gear box design is well illustrated by a story about Thomas Edison, who once won a competition for the production of such a box. Edison's first attempt at production had failed. He and his workers then refined the wheels and changed to new types of oil, but nothing seemed to help. After long experimentation (the necessity of which surprised Edison, who had vastly underestimated the difficulties inherent in this problem) they were ready to give up. Then, by chance, someone brought in some animal fat. With this, the gear box functioned properly. Edison immediately ordered the box closed, and decided to avoid such projects in the future.

Complexity is not confined to industrial processes, however. Consider, for instance, the operation of a computer hard drive. Here, the read/write head must be maintained within a few microns of the spinning magnetic disk at all times. This cannot be done using levers or other mechanical parts; it must be accomplished through a complex design utilizing airflow between the parts. A slightly more mundane but no less interesting example is the problem of knife sharpening. Success here will require much more than the mere maintenance of a good angle between the blade and some randomly chosen stone. In order to function at all, the surface of a sharpening stone must actually crumble during use; this allows for subtle but necessary changes in angle with respect to the knife blade. Some stones even react chemically with the blade, reinforcing the crystal structure of the steel near the sharpened edge. Razor blades turn out to be even more interesting. A microscope will reveal a dual-edged structure whose elastic behavior makes possible a clean shave.

We come across many processes whose complexity we fail to appreciate. How, for example, does a tall tree manage to transport hundreds of liters of water per day from its roots to its leaves? It was shown long ago that a simple vacuum pump cannot raise a water column to a height of more than about ten meters. Capillary action seems unable to account for the phenomenon as well. A tree has no moving parts! This is one of many open problems that still challenge researchers. Others, of course, include the occurrence of catastrophes such as that involving the space shuttle *Challenger*.

Let us finish our brief consideration of complexity with a few examples of complex things that can be explained with simple arguments.

4.5 The Role of the Cosine in Everyday Measurements

We know that the deepest of thinkers put forth tremendous efforts in order to develop a precise definition for the length of a curve and methods for its calculation. The fact that we seem to be able to measure the circumference of our waistlines easily and accurately (sometimes more accurately than we would like) is really no indication of the difficulties inherent in the problem of precise measurement. A tape measure cannot be made to exactly coincide with an arbitrary curve. Furthermore, several successive measurements made along the same curve or surface with an ordinary tape measure will yield varying results: small differences in positioning, and especially in tension, contribute to these variations. However, we can expect a fair degree of accuracy even in the presence of such deformations. Why? The reason has to do with the behavior of the cosine function for small angles.

If we zoom in on a tiny portion of a curve and a tape being used to measure it, we find that both can be approximated by straight segments: the latter inclined at a small angle with respect to the former. How do the lengths of these segments differ? Let them have a common corner point, and call the small inclination angle α (measured in radians). From elementary geometry, we know that the biggest ratio between the lengths of the segments occurs when they participate in the formation of a right triangle: that is, when one segment forms the hypotenuse c and the other forms a leg a of the triangle. Then their ratio is $a/c = \cos\alpha$. How close is this ratio to unity? To answer this, we use the Taylor expansion of $\cos\alpha$ at zero:

$$\cos\alpha = 1 - \frac{\alpha^2}{2!} + \frac{\alpha^4}{4!} - \cdots.$$

For small α, we can truncate the series at any term and get an error smaller than the first neglected term. So the first two terms approximate the value of $\cos\alpha$ to within an error less than $\alpha^4/4!$. A better understanding can be had through reference to some actual numbers. Let L be the length of a small arc of a circle of radius r, and d the length of the chord connecting the ends of the arc. We know that $L/d \to 1$ when $d \to 0$. So, for a small arc, the corresponding angle in radians, which is L/r, is well approximated by d/r. If we take a triangle with hypotenuse of unit length and leg opposite to α of length 0.1, we get $\alpha \approx 0.1$; hence the two-term Taylor approximation has an inaccuracy of no more than $0.1^4/4! \approx 0.000004$. This is so good that we can consider the two-term approximation as practically exact. In fact, however, for the same triangle, we have $\alpha^2/2! \approx 0.005$, and this is good accuracy as well (at least where a tape measure is concerned). Furthermore, a realistic value of α might be even smaller — around 0.01, say — and then $\alpha^2/2!$ becomes about 0.00005. When we sum all the small lengths on the tape measure and compare with the real length of the curve, we get an error of the order of the squared difference in angle along the curve. The actual error could be much less because cancellation may occur among individual errors having opposite signs. So we can measure lengths fairly accurately.

The cosine function also plays a role in determining how well we can weigh an object using a beam balance. A good balance will be a well-made, precision instrument; however, it is still true that when we place the unknown mass and calibrated weights in the pans, the beam will bend slightly. More importantly, this bending will not be fully symmetric. But the resulting difference in the moment arms will be small, and the balance will work well anyway.

The same small-argument behavior of $\cos\alpha$ results in the fact that during a wind storm, when a tall building may begin to sway back and forth significantly, the vertical motion of the building will remain almost negligible.

The corresponding small difference between the leg and hypotenuse of a right triangle is used by mechanical engineers who employ the method of virtual displacements. We calculate possible displacements due to rotation through an infinitesimally small angle α, using the initial length, and replacing $\cos\alpha$ by 1 and $\sin\alpha$ by α.

4.6 Accuracy and Precision

We have said that certain mathematical notions must remain undefined: these include the point, straight line, and plane. The situation is similar with mechanics, where mass is taken as a primitive quantity. The notion of mass is needed to formulate two general laws: Newton's second law, and the law of gravitation. But we must be somewhat cautious: there is no a priori reason to expect that these two laws must involve the "same kind" of mass (i.e., that a body's resistance to acceleration must coincide with its ability to attract other bodies). Although this coincidence was suspected, it had to be verified through systematic experimentation with a torsional pendulum. We now know that the *gravitational* and *inertial masses* of an object stand in numerical agreement up to twelve significant decimal places.

We live in an age where extremely precise measurements seem commonplace. Such diverse quantities as electrical resistances, time intervals, and distances to orbiting satellites can be determined with complete confidence. The electron microscope even lets us measure interatomic distances. But there is one important exception to all this: the involvement of mechanical tools in our measurements can lead to marked disappointment. The weight of an object, for example, can be determined to no more than five decimal places, and even this cannot be accomplished without taking into account both temperature and the buoyant force acting on the object.

The precision of mechanical measurements is restricted by many factors, including the accuracy with which the tools were manufactured and various environmental effects such as temperature changes. Even if we aim to fabricate a simple steel cylinder with great precision, we will likely run into a host of problems. First, the lathe will be nonideal: the clamping device it uses to grip the raw material may (due to inaccuracies in its production) rotate with some eccentricity, for example. Our cutting tools may deform slightly even as we use them, both because of the pressure we must exert and the temperature changes associated with friction. More insidious is the fact that we will eventually work our raw material into a "cylinder" whose diameter will keep changing nonuniformly, due to temperature effects, as we move our shaping tool back and forth along its length — thus, we may end up producing a nice surface of revolution but not necessarily a right circular cylinder. Some decades ago, an expensive lathe was put into production. The frame and other parts were initially cast free of prestrain and then put aside for years to allow for final relaxation of the metal. They then underwent a second stage of shape refinement, after which relaxation was again permitted to take place. After that, the last details were added and the lathe was assembled. But rapid, mass production of such a machine was not feasible. Today, the goal is reached more easily and inexpensively through microprocessor-based control of the cutting tool. Of course, the processor code should be based on a good mathematical model for how metals behave under cutting processes. The fact remains, however, that the machining of parts still cannot be done with the kind of accuracy possible in the other areas we mentioned above.

This being said, we must add that the quality of modern mechanical systems depends crucially on the precision of parts manufacture. The performance of an engine, for example, will hinge upon the use of precision bearings. The eccentricity of the bearing balls used in aviation must fall within a tolerance of $0.25\ \mu$, and the associated rings are made to within $1\ \mu$. Bearings subject to a looser tolerance, of $2\ \mu$, say, can greatly reduce the service time of a jet engine. On the other hand, only costly manufacturing processes involving much technological sophistication can keep the tolerances low: a facility to implement a tolerance of $0.25\ \mu$ may take decades to prepare. Even the body heat of a nearby worker can disturb the delicate environmental conditions that must be maintained. Nevertheless, the payoff is always a longer service life of the final product, whether this is a washing machine or a car. Again, microcomputer control will play a large role in the success of the endeavor. We indicated above that the computer code should be based on a solid mathematical model of the physical process. Often, however, it is found that a semiempirical model works best.

In many circumstances, precise measurements are not possible. The omnipresence of factors such as dust may be enough to spoil our prospects. The mode in which a measurement device is operated is also a key issue. Consider, for example, the kinds of scales used to weigh highway vehicles. Such a scale might accurately determine the weight of a stationary truck to within ten kilograms. However, suppose we decide to use this same scale on a moving truck: we agree to obtain the total weight by adding the separate results obtained for each axle. In this case, we might find ourselves off by as much as a ton. Dynamical phenomena such as nearly imperceptible oscillations are often responsible for this and similar problems with other measuring systems. Precise measurements, as a rule, must be found using slow processes. We can speed things up only if we implement computer software capable of correcting for dynamical effects. This is often difficult to do, however. In the case of the truck scale, a tiny stone in the road will be hard to predict in advance! The same is true of small eccentricities in the wheels of the truck, and many other potential sources of inaccuracy.

4.7 How Trees Stand Up against the Wind

We would like to suggest a model that may help explain why trees display such great strength.

We have all seen trees bend in the wind. The drag forces produced by air moving past a large tree crown can be substantial. We have read accounts of how the masts of old ships would snap off during storms at sea. Still, it is relatively rare to see a living tree broken by even a very strong wind. How can a living tree resist the wind with so much more success than a dead wooden mast? What kind of structural integrity is lost when a tree dies? Nature is the greatest engineer, and has managed to solve many complex optimization problems.

We know that the simple bending of an elastic beam by an applied moment is well approximated by Bernoulli's model. The beam has a neutral axis (this would be a line through the centers of the tree cross section in our case) that bends without a change in length. All the cross sections, after bending, remain flat and perpendicular to the neutral axis. So when the beam is bent, the particles on one side of the neutral axis are placed under tension while those on the other side are placed under compression; each of these forces is proportional to the coordinate z measured perpendicularly away from the neutral axis. Many of the beams used in engineering closely obey this model. So it is clear that the external points of the beam (or tree) must play a large role in our discussion; the resistance couple produced by the beam receives its largest contributions from those points having the largest moment-arm coordinates $|z|$. This is why engineers build structures using hollow-pipe members; all the points are under the same state of stress, and material is thereby economized.[3] We do find many grasses constructed in this fashion. Trees, however, are a different story. The difference seems to lie in the fact that a tree trunk has a heavy crown to support. (There may be biological reasons as well, but we will confine ourselves to mechanical considerations.) The vertical strength of a tree is great, with a substantial safety margin. It would be strange if Nature had not provided trees with a degree of bending strength over and above that found in an ordinary dead beam. In the following paragraphs, we shall spend some time examining this issue. Several questions remain open, however, and the subject could certainly benefit from further investigation.

We know that freshly cut wood can shrink and warp as it cures. The loss of internal water brings a loss of internal stresses and their associated strains. It turns out that a living tree trunk is not like a simple concrete or steel beam; rather, it is *prestressed*, with certain parts living under compression and others under tension. This explains the deformation that can occur when the cells lose their water content later. A well-cured wooden board will behave in accordance with Bernoulli's model, but a living tree trunk

---

[3]Of course, if the beam will be subjected to bending in arbitrary planes, then a circular ring-type cross section will ensure equal resistance to bending in any direction.

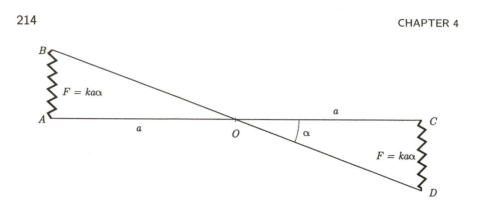

Figure 4.1 Simple model for the bending of a tree trunk.

will not. Although we shall not attempt to discuss the complete theory of tree bending (for which we would at least have to know the prestress distribution in detail), we shall present a rough but adequate qualitative model capable of explaining the principal effect. Our assumption, which is supported by what we see upon cutting through a tree trunk, will be that in its normal state the trunk exists under compression near its center and under tension near its outer boundary.

The wood of the trunk obeys Hooke's linear law. Simple prestressing cannot offer any advantages with regard to bending resistance, because for a linear system, prestressing would not bring anything new into the picture of additional deformation. This means that the bending of a real trunk should involve nonlinearity in the picture of deformation. Such nonlinearity is, in turn, implied by inhomogeneities in the compression of the middle part of the trunk in the normal state. When the trunk is bent and its cross section turns through a small angle α,[4] the point about which the rotation occurred (and so the neutral axis) shifts toward the side that for a non-prestressed round beam would be compressed. Let this shift be equal to δ. In this way, the length of the new neutral axis may be less than that of the initial one, and the reduced tension in the external part of the trunk permits higher values of stretching in that region due to bending. We are interested only in the additional moment that the trunk produces against the rotation of one of its cross sections through a small angle α.

We first present a schematic model of the bending process in the absence of prestress (Figure 4.1). We simulate the reaction of a beam against rotation through an angle α using two elastic springs AB and CD that are clamped to a rigid rotating line BD and to an immovable line AC. The forces asserted by the springs due to rotation about O are $ka\alpha$, where k is the stiffness factor of a spring. Because the moment arm is a for both forces, the total

---

[4]Here, the rotation occurs as though the point of the neutral axis were a point of contact of two circles: when one part inclines, the point of contact moves toward the side of inclination.

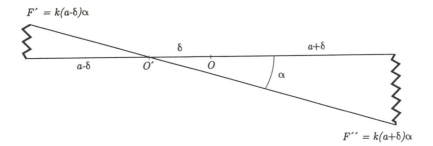

Figure 4.2 Revised model for the bending of a tree trunk.

moment is

$$M = 2k\alpha a^2.$$

Now, suppose the center of rotation shifts a distance δ to the left, as shown in Figure 4.2, because of the inclination angle α associated with the prestressed state of the trunk. The force asserted by spring AB is now $k(a - \delta)\alpha$, and its moment arm for a new position O' of the "neutral axis" is $a - \delta$, so the moment about O' of this force is $k\alpha(a - \delta)^2$. In a similar way, the moment is $k\alpha(a + \delta)^2$ for spring CD. Thus, the moment of the springs rotating about O' through the same angle is their sum:

$$M' = 2k\alpha(a^2 + \delta^2).$$

The increase in moment — in comparison with M for the non-prestressed state — is due to the shift in the neutral axis. Of course, the stress in spring CD increases, but what is important for the safety of the structure is the sum of this additional stress and the prestress. The neutral axis shortens, as we have stated, so the level of prestress is lower. Thus, the additional stress in CD can be higher than for the immovable center of rotation O. Hence, a structure of this type has greater bending rigidity, as is the case with trees. Of course, to calculate the shift of the neutral axis (and its decrease in length) it is necessary to know the distribution of strains in the undeformed state of the tree. Knowledge of this would allow us to calculate the reaction of tree trunks against bending.

In a strong wind, the persons sailing a yacht will reduce the surface area of the sails, and in a dangerous storm, they will drop the sails completely. A tree can do essentially the same thing: in a strong wind, its leaves will assume positions that tend to minimize the drag force on the crown, and it can even drop leaves if necessary. It seems that Nature has optimized the strength of the leaf attachment; however, the shape and size of the attachment region varies from species to species, so it would be interesting to look further into the optimization criterion that was used. The answers to such questions often provide ideas for civil engineers and other design professionals.

4.8 Why King Kong Cannot Be as Terrible as in the Movies

Children like to imagine what would happen if the size of an ant could be increased a thousandfold. So did the producers of the movie about King Kong.

But the problem of altering the size of an object is also central to technical experimentation; here, adult men and women, engineers and researchers, must "scale" the sizes of airplanes, ships, buildings, and so forth. At first glance, this seems both sensible and easy: from tests run on a scale model of an object, we should be able to learn anything we wish to know about the object itself. Things turn out to be a bit more complicated than that. A scale model of an object can exhibit properties that differ markedly from those of the original object. This is a consequence of the fact that our space has three dimensions.

What happens to a cube when we reduce the length of its edges by a factor of 10? All properties of the cube do *not* experience a tenfold reduction: the face area is reduced by a factor of $10 \times 10 = 100$. The volume is reduced by a factor of 1000, as is the weight of the cube, because the weight is proportional to the volume.

Generally speaking, when we reduce the linear dimensions of an object by some factor, we reduce its volume by the cube of that same factor. Some parents know how to take advantage of this. When trying to coax a child into drinking a foul-smelling medicine, they pour the liquid into a cone-shaped liqueur glass and say, "Please drink only half!" The unsuspecting boy or girl drinks until the height of the liquid is reduced by half, but what remains in the conical glass is only one-eighth the original volume. This is a useful application of knowledge; according to Francis Bacon, founder of the modern scientific method, knowledge is power!

Let us consider what happens to a scale model, during simulation, from the viewpoint of strength. The strength of materials considers such problems in detail, but we can make some elementary observations on our own. We know that the resistance a body offers to a force depends on the pressure exerted by the force. We cannot hammer an iron cylinder of 5 cm diameter into a piece of wood, but a 3 mm nail will enter the wood with ease. Indeed, the action of one body on another is determined not only by the interaction force but also by the area S over which this force acts. We characterize this effect by a quantity we call pressure (or normal stress in the strength of materials) and define it using the equation $p = F/S$. A body can resist a small pressure even if the total force is large, but it cannot resist a small force that corresponds to a large pressure (the case of the nail mentioned above). The ability of a body to offer resistance to a force is determined by certain critical values of applied pressure. The larger the application area, the larger the force that can be applied without destroying a body.

From this same viewpoint, let us consider what happens to a wooden cube when we decrease its edge length from 1 m to 10 cm. The mass of a 1 m cube is about 700 kg, so its weight is about 700 kg $\times$ 9.8 m/s^2 $\approx$ 7000 N.

Its bottom face has area 1 m², so the pressure exerted by the cube against the floor is about 7000 Pa. On the other hand, a 10 cm cube has mass 0.7 kg and face area 0.01 m², and therefore exerts a pressure of about 700 Pa. We see that the bigger cube exerts ten times the pressure exerted by the smaller cube. Only the stresses (pressures) are important when we consider the question of strength of an item. The importance of this will become apparent momentarily.

Let us calculate the stress in a person's tibia (major leg bone below the knee) when that person carries a 60 kg bag of potatoes on his shoulders. Suppose the person himself weighs 60 kg as well, and that the cross-sectional area of the bone is 6 cm². We wish only to obtain a rough estimate.[5] Thus, the total cross-sectional area of the two tibia bones is 12 cm², and the pressure from the bag is 60 kg × 10 m/s² ÷ 0.0012 m² = 50000 Pa. The pressure from the person's own body weight is roughly the same value: 50000 Pa. Now, let us suppose that everything, the person and the bag, could be scaled down in linear dimensions by a factor of 100. From what we said in the paragraph above, the pressure should decrease by a factor of 100 as well. Let us also assume that the person's reduced body is still composed of the same materials as before; that is, the material of the reduced tibia bones can withstand roughly the same pressure as before. Then the reduced person can move and exist under the old pressures, which totaled 100000 Pa. But the decreased pressure from his own weight and the reduced bag is only 1/100 of that total. This means that we can safely increase the load on the man 99-fold, that is, we can add 198 more bags of reduced potatoes. We see that in doing calculations we can simply forget about the weight of the reduced man, which was impossible before our hypothetical size reduction occurred.

Let us now turn this situation around. An imaginative child who sees an ant carry around ten times its own weight might believe that this would still be possible if the ant were to become many times larger. If an ant could reach elephant size, could it hoist and carry ten elephants with ease? Probably not, and it is very possible that such an unnaturally large ant would collapse under its own weight. If all bodily dimensions of an ant were to increase 100-fold, then all cross-sectional areas would increase 10,000-fold, and the weights of all body parts would increase 1,000,000-fold; therefore, all pressures exerted under the weights of those body parts would increase 100-fold (an ant has no bones, but the situation with its exterior supportive framework — its *exoskeleton* — is similar).

This difference in weight and pressure in cross sections during scaling is a consequence of the laws of mechanics, and so it is the same for any creature.

---

[5]Here we discuss the human body as if it were as homogeneous as a piece of wood. We neglect many details like the weight of the feet, the fact that there is also a fibula bone in the same leg, that the pressure is not distributed uniformly throughout the bone, and so on. The result will be a rough model which may fail in certain particulars but that should give us a fairly good general picture of the situation. Even professional researchers do not strive for absolute accuracy in their preliminary calculations.

A 100-fold magnification of a monkey would produce a creature who would experience a 100-fold increase in the pressure in the supportive tissues of his feet. Such a "King Kong" would probably have to slither around on all fours, and how frightening would that be?

The situation with proportional changes in weight, pressure, strength, and so forth, results in the fact that when a small insect moves, the stresses in its skeleton under its own weight are small compared to those associated with weight it can pick up or push. As the size of an animal increases, so does the load of its own body on its supportive framework; when some limiting size is reached, the strength of that framework will go purely to support the body. If we were given the chance to design an animal body using some set of elements — bones, muscles, and so on — we would be quite restricted in terms of maximum possible size. If we kept to less than the critical size, we could let various other criteria influence our design. We could, for example, design for running speed (we would need to introduce a horizontal position for running because it is more economical than the vertical position humans use). A variety of such conditions could determine the various forms and sizes of creatures that we might dream up.

So the King Kong of the movies is terrible, but a real-life version could probably manage to duplicate only the loud roars and sighs of its famous namesake.

Afterword

Mechanics is among the oldest of the sciences. Its development was contingent on the tools of mathematics; in turn, a great deal of mathematical activity was stimulated by the necessity to solve problems from mechanics and related engineering disciplines. We have considered some of the ideas that form the groundwork of mechanics. In doing so, we have seen that many natural phenomena can be described through the use of mechanical tools and models. At the same time, it has become evident that true understanding cannot be reached through the use of a mere "formula plugging" approach: those who wish to make substantial contributions to science and engineering must go beyond the formulas that have come to dominate modern textbooks. It is especially the case that an unlimited, almost blind use of computers will continue to bring catastrophic results.

Similar things can be said regarding almost any quantitative science, so students in practically any area can gain from the kind of approach we have taken in this book. The connection between any science and its mathematical roots is a particularly instructive thing to examine, and workers in all disciplines can learn much from the kind of consideration we have given to mechanics and its foundations.

Recommended Reading

In this book we often mentioned "more detailed sources." Of course, many good textbooks are appropriate for this. We list only a few of these here.

The problems of calculus, including limits, series, derivatives, and integrals, can be found in

Marsden J.E. and Weinstein A. (1990). *Calculus* (three volumes). Springer–Verlag.

The main tools of theoretical mechanics can be found in

Lurie A.I. (2002). *Analytical Mechanics.* Springer–Verlag.

The strength of materials is presented in

Timoshenko S. (1968). *Elements of Strength of Materials.* Wadsworth Publishing Company.

A book by the same author can be recommended as a simple introduction to the theory of elasticity:

Timoshenko S. and Goodier J. (1970). *Theory of Elasticity.* McGraw Hill.

The classic text on hydrodynamics is

Lamb H. (1993). *Hydrodynamics.* Cambridge University Press.

Among the many good texts on ordinary differential equations, we could cite

Boyce W. and DiPrima R. (2000). *Elementary Differential Equations and Boundary Value Problems.* Wiley.

For partial differential equations the reader could consult

Powers D. (1987). *Boundary Value Problems.* International Thomson Publishing.

The theory of tensors, on an elementary level, can be found in

Lebedev L.P. and Cloud M.J. (2003). *Tensor Analysis.* World Scientific.

Finally, the history of mathematics is presented in

Kline M. (1990). *Mathematical Thought From Ancient to Modern Times.* Oxford University Press.

Index